METHODS IN MOLECULAR BIOLOGY™

Series Editor
John M. Walker
School of Life Sciences
University of Hertfordshire
Hatfield, Hertfordshire, AL10 9AB, UK

For further volumes:
http://www.springer.com/series/7651

Membrane Proteins

Folding, Association, and Design

Edited by

Giovanna Ghirlanda

Department of Chemistry and Biochemistry, Arizona State University, Tempe, AZ, USA

Alessandro Senes

Department of Biochemistry, University of Wisconsin-Madison, Madison, WI, USA

Humana Press

Editors
Giovanna Ghirlanda
Department of Chemistry and Biochemistry
Arizona State University
Tempe, AZ, USA

Alessandro Senes
Department of Biochemistry
University of Wisconsin-Madison
Madison, WI, USA

ISSN 1064-3745   ISSN 1940-6029 (electronic)
ISBN 978-1-62703-582-8   ISBN 978-1-62703-583-5 (eBook)
DOI 10.1007/978-1-62703-583-5
Springer New York Heidelberg Dordrecht London

Library of Congress Control Number: 2013944656

Printed on acid-free paper

Humana Press is a brand of Springer
Springer is part of Springer Science+Business Media (www.springer.com)

Preface

While the importance of membrane proteins as gateways to the cell has long been recognized—for example, membrane proteins account for a majority of current drug targets—challenges associated with experimental and structural characterization have delayed the development of dedicated engineering methods. However, recent advances in our understanding of folding and function of membrane proteins have opened up the tantalizing possibility of designing novel membrane proteins with tailored functionality.

In this volume of *Methods in Molecular Biology*, we focus on model systems for the study of structure, folding, and association in the membrane, and present an overview of methods that can be applied to these systems. Most work so far has focused on model peptides, which have been used extensively to establish current paradigms for protein folding in the membrane, as well as to identify sequence motifs that mediate protein–protein interactions. Mirroring the trajectory followed for water-soluble model systems, the field is now rapidly shifting from using design as a tool to dissect critical features of natural proteins to the design of functional artificial proteins.

The book is divided into four parts. The first part groups articles that cover a wide variety of methods for measuring the association of transmembrane helices. The first three chapters provide detailed protocols for a series of complementary biophysical methods for measuring association of transmembrane peptides in lipids and detergents, namely, thiol-disulfide exchange (Chap. 1, Cristian and Zhang), FRET (Chap. 2, Khadria and Senes), and a recently developed method based on steric trapping (Chap. 3, Hong et al.). In Chap. 4, Schneider et al. present a comprehensive discussion of in vivo reporter systems for assessing transmembrane interactions in biological membranes.

The interaction of transmembrane proteins with the lipid bilayer is one of the most important factors that contribute to their folding. Chapter 5 (Caputo) presents a protocol for the study of the insertion of transmembrane helices in synthetic liposomes by fluorescence. Chapter 6 (Findlay and Booth) presents a protocol for folding membrane proteins in lipid bilayers. Chapter 7 (Lomize and Pogozheva) describes a solvation potential that can be used to predict computationally the orientation of proteins in membranes.

Two structural chapters cover the characterization of membrane proteins by NMR. Chapter 8 (Marassi et al.) discusses solid-state methods for the study of membrane proteins in lipid bilayers. Chapter 9 (Veglia et al.) presents NMR methods for the study of the interactions of antimicrobial peptides with lipids.

The final part groups a series of articles that share engineering as a common theme. In Chap. 10, Nanda et al. discuss computational methods for the prediction and design of beta-barrel transmembrane proteins. Rath and Deber present a protocol for the rational design and synthesis of transmembrane peptides (Chap. 11). Lluis and Yin (Chap. 12) discuss the engineering of a reporter cell line for studying receptor–peptide interactions. Finally, in Chap. 13, Krishnamurthy and Kumar discuss the use of fluorination for the design of transmembrane assemblies of peptides.

Tempe, AZ, USA   *Giovanna Ghirlanda*
Madison, WI, USA   *Alessandro Senes*

Contents

Contributors

GEORGE BARANY • *Department of Chemistry, University of Minnesota, Minneapolis, MN, USA*

PAULA J. BOOTH • *Department of Biochemistry, School of Medical Sciences, University of Bristol, Bristol, UK*

JAMES U. BOWIE • *Department of Chemistry and Biochemistry, UCLA-DOE Institute for Genomics and Proteomics, Molecular Biology Institute, University of California, Los Angeles, CA, USA*

GREGORY A. CAPUTO • *Department of Chemistry and Biochemistry, Rowan University, Glassboro, NJ, USA*

YU-CHU CHANG • *Department of Chemistry and Biochemistry, UCLA-DOE Institute for Genomics and Proteomics, Molecular Biology Institute, University of California, Los Angeles, CA, USA*

LIDIA CRISTIAN • *Influmedix, Inc., Radnor, PA, USA*

ALEXANDER DAVIS • *Department of Biochemistry and Molecular Biology, Center for Advanced Biotechnology and Medicine, Robert Wood Johnson Medical School—UMDNJ, Piscataway, NJ, USA*

CHARLES M. DEBER • *Division of Molecular Structure & Function, Research Institute, Hospital for Sick Children, Toronto, ON, Canada*

YI DING • *Sanford Burnham Medical Research Institute, La Jolla, CA, USA*

HEATHER E. FINDLAY • *Department of Biochemistry, School of Medical Sciences, University of Bristol, Bristol, UK*

HEEDEOK HONG • *Department of Chemistry, Michigan State University, East Lansing, MI, USA; Department of Biochemistry & Molecular Biology, Michigan State University, East Lansing, MI, USA*

DANIEL HSIEH • *Department of Biochemistry and Molecular Biology, Center for Advanced Biotechnology and Medicine, Robert Wood Johnson Medical School—UMDNJ, Piscataway, NJ, USA*

AMBALIKA KHADRIA • *Department of Biochemistry, University of Wisconsin-Madison, Madison, WI, USA*

VIJAY M. KRISHNAMURTHY • *Department of Chemistry, Tufts University, Medford, MA, USA*

KRISHNA KUMAR • *Department of Chemistry, Tufts University, Medford, MA, USA; Cancer Center, Tufts Medical Center, Boston, MA, USA*

MATTHEW W. LLUIS • *Department of Chemistry and Biochemistry and the BioFrontiers Institute, University of Colorado at Boulder, Boulder, CO, USA*

ANDREI L. LOMIZE • *Department of Medicinal Chemistry, College of Pharmacy, University of Michigan, Ann Arbor, MI, USA*

FRANCESCA M. MARASSI • *Sanford Burnham Medical Research Institute, La Jolla, CA, USA*

VIKAS NANDA • *Department of Biochemistry and Molecular Biology, Robert Wood Johnson Medical School—UMDNJ, Piscataway, NJ, USA*

STANLEY J. OPELLA • *Department of Chemistry and Biochemistry, University of California, San Diego, La Jolla, CA, USA*

IRINA D. POGOZHEVA • *Department of Medicinal Chemistry, College of Pharmacy, University of Michigan, Ann Arbor, MI, USA*

FERNANDO PORCELLI • *Department of Chemistry, University of Minnesota, Minneapolis, MN, USA; Department of Biochemistry, Molecular Biology, & Biophysics, University of Minnesota, Minneapolis, MN, USA*

AYYALUSAMY RAMAMOORTHY • *Department of Chemistry, University of Michigan, Ann Arbor, MI, USA*

ARIANNA RATH • *Division of Molecular Structure & Function, Research Institute, Hospital for Sick Children, Toronto, ON, Canada*

DIRK SCHNEIDER • *Department of Pharmacy and Biochemistry, Johannes Gutenberg University, Mainz, Germany*

ALESSANDRO SENES • *Department of Biochemistry, University of Wisconsin-Madison, Madison, WI, USA*

DOMINIK STEINDORF • *Department of Pharmacy and Biochemistry, Johannes Gutenberg University, Mainz, Germany*

YE TIAN • *Sanford Burnham Medical Research Institute, La Jolla, CA, USA; Department of Chemistry and Biochemistry, University of California, San Diego, La Jolla, CA, USA*

LYDIA TOME • *Department of Pharmacy and Biochemistry, Johannes Gutenberg University, Mainz, Germany*

GIANLUIGI VEGLIA • *Department of Chemistry, University of Minnesota, Minneapolis, MN, USA; Department of Biochemistry, Molecular Biology, & Biophysics, University of Minnesota, Minneapolis, MN, USA*

YONG YAO • *Sanford Burnham Medical Research Institute, La Jolla, CA, USA*

HANG YIN • *Department of Chemistry and Biochemistry and the BioFrontiers Institute, University of Colorado at Boulder, Boulder, CO, USA*

YAO ZHANG • *GlaxoSmithKlineplc, King of Prussia, PA, USA*

Part I

Association of Transmembrane Helices

Chapter 1

Use of Thiol-Disulfide Exchange Method to Study Transmembrane Peptide Association in Membrane Environments

Lidia Cristian and Yao Zhang

Abstract

The development of methods for reversibly folding membrane proteins in a two-state manner remains a considerable challenge for studies of membrane protein stability. In recent years, a variety of techniques have been established and studies of membrane protein folding thermodynamics in the native bilayer environments have become feasible. Here we present the thiol-disulfide exchange method, a promising experimental approach for investigating the thermodynamics of transmembrane (TM) helix–helix association in membrane-mimicking environments. The method involves initiating disulfide cross-linking of a protein under reversible redox conditions in a thiol-disulfide buffer and quantitative assessment of the extent of cross-linking at equilibrium. This experimental method provides a broadly applicable tool for thermodynamic studies of folding, oligomerization, and helix–helix interactions of membrane proteins.

Key words Transmembrane helix, Association, Equilibrium, Thermodynamics, Energetics, Oligomerization, Topology, Thiol-disulfide exchange

1 Introduction

One of the greatest challenges in measuring the thermodynamics of membrane protein folding is finding conditions under which the proteins fold reversibly in a two-state process between the native, folded state and a well-defined unfolded state. In contrast to water-soluble proteins, where the unfolded state largely lacks secondary structure, the unfolded state of helical membrane proteins retains considerable helical content, thus presenting additional challenges in quantitatively measuring the association energetics. In recent years, various biophysical tools have been developed to measure TM helix–helix interactions in membrane-mimicking environments, including analytical ultracentrifugation (AUC), fluorescence resonance energy transfer (FRET), and thiol-disulfide exchange [1–6].

Giovanna Ghirlanda and Alessandro Senes (eds.), *Membrane Proteins: Folding, Association, and Design*,
Methods in Molecular Biology, vol. 1063, DOI 10.1007/978-1-62703-583-5_1, © Springer Science+Business Media, LLC 2013

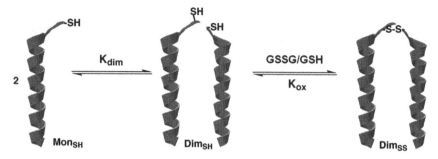

Fig. 1 The principle of measuring a reversible association reaction by disulfide cross-linking involves initiating disulfide bonds of a protein under reversible redox conditions using a glutathione redox buffer (oxidized: GSSG and reduced: GSH) and quantitative assessment of the extent of cross-linking at equilibrium

The thiol-disulfide exchange method relies on proximity as a measure of interaction between TM helices and can be applied in either detergent micelle or lipid bilayer environments. It requires the introduction of a cysteine label in the studied system (if a natural one is unavailable) and relies on the formation of disulfide bonds between cysteine on different protein subunits at redox potentials set by varying the ratio of reduced to oxidized redox couple (e.g., glutathione) (Fig. 1). The method yields mole fractions of disulfide cross-linked dimers, based on which the association affinity can be calculated. In addition, the method can be used to determine the orientational preference (topology) of TM helices. This method also applies to oligomeric orders higher than dimer, where the fraction of oligomer has to be inferred from the fraction of dimer (see examples in Subheading 4).

2   Materials

2.1 Incorporation of Peptide into Detergent Micelles or Lipid Vesicles

1. Lipids and detergents utilized in these studies (and not limited to): dodecylphosphocholine (DPC) (Avanti Polaris Lipids, stored in –20 °C), 1-palmitoyl-2-oleoyl-*sn*-glycero-3-phosphocholine (POPC) (Avanti Polaris Lipids, stored in –20 °C), 1,2-dilauroyl-*sn*-glycero-3-phosphocholine (DLPC) (Avanti Polaris Lipids, stored in –20 °C), 1,2-dimyristoyl-*sn*-glycero-3-phosphocholine (DMPC) (Avanti Polaris Lipids, stored in –20 °C).

2. Trifluoroethanol (TFE).

3. Ethanol.

4. Chloroform.

5. Nitrogen.

6. Vacuum.

7. Tris Buffer: 0.1 M Tris–HCl, 0.2 M KCl, 1 mM EDTA, pH 8.6.

8. Vortex mixer.

2.2 Thiol-Disulfide Exchange Equilibria

1. Redox pair: oxidized (GSSG) and reduced (GSH) glutathione (Sigma-Aldrich).
2. Screw-capped glass vials.
3. Nitrogen.
4. HCl (1.2 M).

2.3 Separation and Identification of the Equilibrium Mixture

1. Reverse-phase (RP) high-performance liquid chromatography (HPLC) instrument (Agilent).
2. C4 analytical HPLC column (Grace Vydac).
3. Separation buffer A (0.1 % TFA, 99.9 % water).
4. Separation buffer B (6:3:1 2-propanol:acetonitrile:water containing 0.1 % TFA).
5. Matrix-assisted laser desorption/ionization-time of fly (MALDI-TOF).

2.4 Data Analysis

1. Igor Pro (WaveMetrics) was utilized as a data analysis program in these studies.

3 Methods

The following general protocol has been utilized for thiol-disulfide exchange studies in detergent micelles or lipid bilayers.

3.1 Incorporate Peptide into Detergent Micelles or Lipid Vesicles

1. Mix the peptide of interest (from an organic solvent stock, generally TFE) with the appropriate detergent or lipid from an ethanol, TFE, or chloroform stock at the desired final peptide/detergent or peptide/phospholipid mole ratio.
2. Remove solvents by evaporation under a stream of nitrogen and dry the resultant films overnight under high vacuum.
3. Hydrate the peptide/lipid/detergent mixed films with a degassed buffer (100 mM Tris–HCl, 0.2 M KCl, 1 mM EDTA, pH 8.6) and disperse samples by vigorous vortexing. In the case of vesicle preparations, sonicate the hydrated peptide/lipid samples to clarity using a bath sonicator.

3.2 Initiate Disulfide Exchange by Addition of a Redox Buffer (Glutathione)

1. Initiate the thiol-disulfide interchange reaction by adding oxidized (GSSG) and reduced (GSH) glutathione at mole ratios of 0.25–0.5 (*see* **Notes 1** and **2**).
2. Vortex samples vigorously and sonicate (in the case of lipid samples) until the peptide/detergent or lipid film is completely incorporated into the glutathione-containing buffer.
3. Conduct reactions under anaerobic conditions to prevent adventitious air oxidation (*see* **Note 3**).

$$GSSG \overbrace{\quad} 2GSH$$
$$2Mon_{SH} \underset{K_{dim}}{\overset{}{\rightleftharpoons}} Dim_{SH} \underset{K_{ox}}{\overset{}{\rightleftharpoons}} Dim_{SS}$$

$$K_{dim} = \frac{[Dim_{SH}]}{[Mon_{SH}]^2} \qquad K_{ox} = \frac{[Dim_{SS}][GSH]^2}{[Dim_{SH}][GSSG]}$$

Scheme 1 The thermodynamic equilibrium between disulfide coupling and dimer formation. (*Abbreviation*: "Mon_{SH}" refers to "Monomer," "Dim_{SH}" refers to "non-cross-linked dimer," and "Dim_{SS}" refers to "cross-linked dimer." "[]" refers to concentration) (Reprinted with permission from [8]. Copyright (2009) American Chemical Society)

3.3 Allow for Equilibrium to be Reached

1. Allow samples to equilibrate for a few hours.
2. A time course analysis by HPLC is recommended in order to make sure that equilibration is reached. In the case of vesicle samples, in order to ensure equilibration of the peptides between vesicles, the samples are subjected to repeated freeze–thaw, followed by sonication until samples are clear.
3. To check reversibility, equilibration reactions can be carried out in the same manner but starting with the oxidized peptide (dimer) incorporated into detergent or lipids and increasing the reducing composition of the redox buffer.

The equilibrium undergoes two steps: an association step and an oxidation step. An equilibrium scheme using dimerization as an example is shown in Scheme 1. The dimerization step (K_{dim}) is a reversible bimolecular association reaction that depends on the reduced monomer concentration (Mon_{SH}); the subsequent oxidation step (K_{ox}) is also reversible but independent of the concentration of the peptide and dependent on the ratio of oxidized (GSSG) to reduced glutathione (GSH). In Subheading 4, a case study of tetramerization will be discussed.

3.4 Quench Reaction by Lowering the pH

After equilibration is reached, the reaction is quenched by lowering the pH with excess HCl (about 100-fold over the glutathione concentration).

3.5 Separate and Quantitate Products by RP-HPLC

1. The products of the thiol-disulfide exchange reactions are analyzed by RP-HPLC. Elutant species generally consist of lipid/detergent, mixed disulfide of peptide with GSH, the thiol-free monomer, and the disulfide-bonded peptide (Fig. 2).
2. The latter three species can be identified by MALDI-TOF mass spectrometry. Assignment of the reduced and oxidized HPLC peaks can also be achieved by HPLC analysis of authentic samples of the oxidized and reduced peptides.
3. A small amount of the reaction mixture is used to determine the total free thiol content at the end of reaction by Ellman test [7].

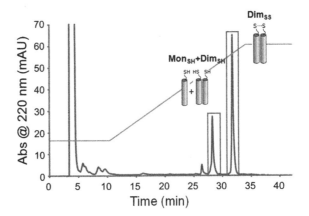

Fig. 2 A typical RP-HPLC chromatogram of an equilibration mixture in a glutathione redox buffer. The peaks labeled Mon_{SH} and Dim_{SS} correspond to the reduced and disulfide-bonded species, respectively

4. Assuming that in the HPLC of the equilibrated mixtures, all species were quantitatively eluted and exhibited identical extinction coefficients at 220 nm (or 280 nm if a significant number of tryptophan residues are present in the peptide sequence). The HPLC peak areas of the present species are integrated using the software supplied with the HPLC instrument.

5. In order to accurately quantify the products of the reaction, *standard concentration curves* for the reduced and oxidized species are recommended and can be prepared as described below.

3.5.1 Preparation of Standard Curves for the Reduced and Oxidized Species

The major experimental observables are the reduced and oxidized species. For example, in the dimerization system described in Scheme 1, RP-HPLC does not distinguish between Mon_{SH} and Dim_{SH} (Fig. 2), therefore the total concentration of the reduced species, $[Pep_{SH}]$, is given by the sum of $[Mon_{SH}]$ and $2[Dim_{SH}]$. Glutathione adducts are also observed, but are generally very low in concentration, and can be neglected because the equilibrium of interest required to compute K_{dim} and K_{ox} involves only the concentration at equilibrium of the reduced species (Mon_{SH} plus Dim_{SH}) as well as Dim_{SS}. In order to quantify each species, standard curves for the reduced and oxidized species are prepared as follows:

1. Standard curves to quantify the reduced species, Pep_{SH}: prepare peptide/lipid (or peptide/detergent) samples (at a wide range of concentrations) and mix each sample with 100-fold (molar ratio) of reduced (GSH) glutathione. After the system reaches equilibrium, separate samples by RP-HPLC. Plot the peak areas of the reduced peptide as a linear function of $[Pep_{SH}]$ to obtain the reduced standard curve.

2. Standard curves to quantify the oxidized species Dim_{SS} are prepared in a similar manner: Peptide samples of increasing peptide/lipid (or detergent) molar ratios are air-oxidized at pH 8.8 overnight and analyzed by RP-HPLC. The chromatogram of each sample will present a reduced-species peak (Pep_{SH}) and an oxidized species peak (Dim_{SS}). [Dim_{SS}] can be calculated by subtracting [Pep_{SH}] from the initial total peptide concentration. Plot the areas of the oxidized peaks as a linear function of [Dim_{SS}], to obtain the oxidized standard curve.

3.6 Data Analysis

1. Based on the equilibrium equations shown in Scheme 1, the material balances for a dimer system can be derived as below:

$$[PT] = [Mon_{SH}] + 2[Dim_{SH}] + 2[Dim_{SS}]; \quad [Pep_{SH}] = [Mon_{SH}] + 2[Dim_{SH}] \tag{1}$$

$$[PT] = [Mon_{SH}] + [Mon_{SH}]^2 K_{dim} + [Mon_{SH}]^2 K_{dim} K_{ox}[GSSG]/[GSH]^2 \tag{2}$$

2. Equation 2 is solved numerically for [Mon_{SH}] as a function of [PT], K_{dim}, K_{ox}, and [GSSG]/[GSH]^2 using root-finding algorithms in IGOR Pro. [PT] and [GSSG]/[GSH]^2 are known, leaving only K_{dim} and K_{ox} as dependent variables. The data is then expressed as a plot of fractions of the peptide in the disulfide form (frac = $2[Dim_{SS}]/[PT]$), and the values of the dependent variables obtained by nonlinear least-squares fitting to the equation: frac = $\{[Mon_{SH}]^2 K_{dim} K_{ox}[GSSG]/[GSH]^2\}/PT + C$ in which C is a constant, which was found to be close to zero for all peptides. This fitting equation is derived from the equilibrium relationships demonstrated in Scheme 1.

This data analysis algorithm can be applied to other oligomerization systems with adequate modifications.

4 Applications

In this section, a few applications of the disulfide-exchange method are presented. The method has been used to study the thermodynamics of TM helix association in detergent micelles [8–10] as well as lipid bilayers [3, 10] or to determine the association energetics and preferred orientation (topology) of TM helices in membrane-mimicking environments [8, 11]. We will begin by citing an example of a model peptide system studied in detergent micelles followed by other systems investigated in lipid bilayers.

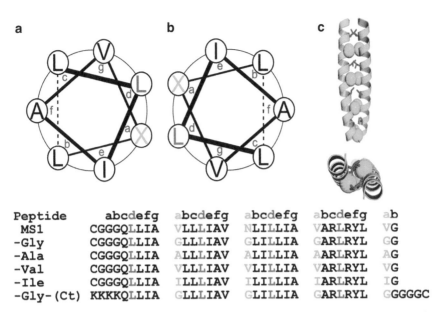

Peptide	abcdefg	abcdefg	abcdefg	abcdefg	ab
MS1	CGGGQLLIA	VLLLIAV	NLILLIA	VARLRYL	VG
-Gly	CGGGQLLIA	GLLLIAV	GLILLIA	GARLRYL	GG
-Ala	CGGGQLLIA	ALLLIAV	ALILLIA	AARLRYL	AG
-Val	CGGGQLLIA	VLLLIAV	VLILLIA	VARLRYL	VG
-Ile	CGGGQLLIA	ILLLIAV	ILILLIA	IARLRYL	IG
-Gly-(Ct)	KKKKQLLIA	GLLLIAV	GLILLIA	GARLRYL	GGGGGC

Fig. 3 Heptad repeats (abcdefg) in helical wheel (**a**), computational model showing side and top view (**b**), and sequence of MS1 variants (**c**). MS1, -Gly, -Ala, -Val, and -Ile are N-terminally Cys-modified. -Gly-(Ct) is C-terminally Cys-modified. The variable "a" positions are shown in *green*, and the Leu at "d" in *red* (Reprinted with permission from [8]. Copyright (2009) American Chemical Society)

4.1 Thiol-Disulfide Exchange in Evaluating TM Association in Detergent Micelles

4.1.1 Energetics of TM Helix Association in Detergent Micelles

In a recent paper [8] that compares the effects of packing large versus small apolar side chains in TM helix association, the method of thiol-disulfide exchange has been employed to successfully identify the association affinity and orientation of TM helices in detergent micelles.

This paper utilized MS1, a membrane-soluble derivative of the coiled coil GCN4-P1 system [12], as a model peptide. Coiled coils represent a ubiquitous and important structural motif in proteins [13]. Coiled coils are characterized by a so called heptad repeat in their primary sequences (abcdefg) [14, 15], in which the "a" and "d" positions are present at the interface and play essential roles in helix association (Fig. 3a). Altering the size of apolar residues in the putative "a" position of the MS1 peptide allows us to investigate how packing in the interaction surface affects TM helix association.

A series of MS1 variants were synthesized in which each of the four "a" positions was varied to Gly, Ala, Val, and Ile (Fig. 3). *These variants were found to be* predominantly helical in DPC micelles over the entire range of peptide/detergent ratios studied in this project, as determined by circular dichroism (CD), while analytical ultracentrifugation (AUC) experiments showed that in DPC micelles, MS1-Gly was fully dimeric, MS1-Ala adopted a monomer–dimer equilibrium, and MS1-Val and MS1-Ile were predominantly monomeric.

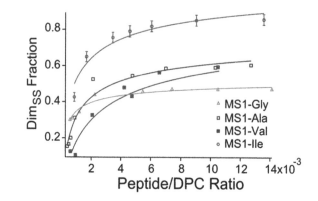

Fig. 4 Analysis of the fraction of cross-linked dimer as a function of peptide/DPC ratios for each MS1/variant. The theoretical curve describes the least-squares fit to Scheme 1. The standard errors in the experimental points are similar for each peptide and are indicated for MS1-Ile (Reprinted with permission from [8]. Copyright (2009) American Chemical Society)

Table 1
pK_{dim} and pK_{ox} obtained from analysis of data in Fig. 4 (Reprinted with permission from [8]. Copyright (2009) American Chemical Society)

	pK_{dim}	pK_{ox}	ΔG_{dim} (kcal/mol dimer)
MS1	−2.6	1.5	−3.6
-Gly	−3.0	2.6	−4.1
-Ala	−1.6	1.2	−2.2
-Val	0.9	−0.8	1.2
-Ile	1.9	−2.0	2.6

The method of thiol-disulfide exchange was then applied to quantify the dimerization (association) affinity of the MS1 variants. Data obtained from these experiments was processed as discussed in Subheading 3. Dim_{SS} fraction for each variant was plotted as a function of peptide/DPC molar ratio (Fig. 4) and fit to determine K_{dim} and K_{ox} for each MS1 variant (Table 1). ΔG_{dim} is then calculated based on K_{dim} using Gibbs free energy function. Comparison of ΔG_{dim} among all the variants suggests that the amino acid in the "a" position aids the association of membrane helices in increasing order of Gly > Ala > Val > Ile, in good agreement with the AUC data and opposite to the ranking observed in water-soluble coiled systems [16, 17].

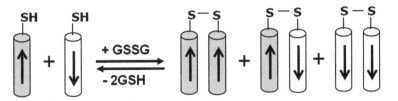

Fig. 5 Schematic diagram of the thiol-disulfide exchange assay used to determine the orientation of the MS1-Gly helices. A mixture of N- and C-terminally Cysteine-labeled peptides that can form both parallel and antiparallel dimers is incubated in a glutathione redox buffer that allows the peptide monomers and disulfide bonds to rearrange (Reproduced from [11] with permission from Elsevier)

4.1.2 Topology of TM Helix Association in Detergent Micelles

The comparison between K_{dim} and K_{ox} in Table 1 reveals that as the size of the side chain at the "a" position decreases, the ease of disulfide formation (reflected in K_{ox}) becomes less favorable. This effect can be seen in Fig. 4: the fraction of disulfide formation levels off at relatively low [Peptide]/[DPC] for MS1-Gly, which reflects high dimerization affinity. However, under these redox conditions, the curve for MS1-Gly extrapolates to a low fraction disulfide at high [Peptide]/[DPC], indicating that disulfide formation is thermodynamically less favorable than for the other variants. These results indicate that MS1-Gly possibly prefers to adopt an antiparallel orientation.

To assess the ability of the MS1-Gly peptides to form parallel versus antiparallel dimers, two peptides were prepared: one with an N-terminal cysteine (MS1-Gly-(Nt)) and the other with a C-terminal cysteine (MS1-Gly-(Ct)), as illustrated in Fig. 3c. Equimolar amounts of the two peptides that can react with themselves as well as each other to form parallel- or antiparallel-stranded dimers were incubated in a glutathione redox buffer and allowed to undergo thiol-disulfide rearrangement (Fig. 5). Depending on the antiparallel or parallel orientational preferences of the peptides, either the heterodimer or the two homodimers will form, respectively, under redox equilibrium conditions. The molar ratio of the disulfide-linked heterodimers versus homodimers can be evaluated by RP-HPLC and MALDI-TOF and the observed molar ratio should reflect the relative stabilities of the parallel and antiparallel disulfide-bonded dimers. For a completely random association of the two cysteine-tagged peptides, the expected statistical distribution of the products would be a 1:1:2 ratio for the two homodimer and heterodimer species, respectively.

A 1:1:14 molar ratio for the two homodimer and heterodimer species was observed in the equilibration mixture of the MS1-Gly-(Ct) and MS1-Gly-(Nt) peptides suggesting that, as expected, MS1-Gly peptide strongly prefers to form antiparallel dimers.

The preference of an antiparallel orientation has also been further confirmed by AUC. The results demonstrate the utility of the thiol-disulfide method to correctly identifying the topology of the MS1 system.

4.2 Thiol-Disulfide Exchange in Evaluating the Association of TM Helices in Lipid Bilayers

4.2.1 Thermodynamics of TM Helix Association in Lipid Bilayers

The method of thiol-disulfide interchange has been implemented to measure the energetics of TM peptide association in phospholipid bilayers as well as the effect of lipid composition on the association of TM peptides. M2 transmembrane (M2TM) peptide from the influenza A virus is discussed here as an example. M2 is a small homotetrameric proton channel, consisting of 97-residue monomers, that acts as a highly selective proton channel and plays an important role in the life cycle of the influenza A virus [18–21].

The protein has two cysteine residues at positions 17 and 19 at the extracellular domain, which form a mixture of covalent dimers and tetramers and which are believed to be nonessential for the function of the protein [19]. The peptide used in the thiol-disulfide studies described previously comprised residues 19–46 and contained a single cysteine residue at position 19 which reversibly forms intermolecular disulfides in the tetramer [9].

The protein is known to exist in a monomer/tetramer equilibrium which can be manipulated by adjusting the peptide/phospholipid ratio. If the experiments are conducted under conditions in which the protein is primarily monomeric in the reduced form, but tetrameric when oxidized, one can analyze the peptide concentration dependence of the equilibrium to obtain the dissociation constant for tetramerization (Scheme 2). This method requires a substantial monomer population in equilibrium with the tetrameric form of the protein at experimentally accessible peptide to lipid ratios. Previously, it was found that this could be accomplished by incorporating the M2 peptide into short-chain lipid bilayers formed from DLPC, in which the tetrameric conformation is less stable than in longer chain lipid bilayers such as DMPC or POPC [3]. We used the protocol described in Subheading 3 to carry out the thiol-disulfide experiments in DLPC bilayers and to quantitate

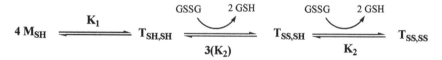

Scheme 2 The thermodynamic equilibrium between disulfide coupling and tetramer formation. (*Abbreviation*: The monomeric and tetrameric states are given as M and T, respectively, and their oxidation states "SH" for reduced and "SS" for oxidized are indicated as a subscript. The factor of 3 for the upper value of K2 is related to the fact that a given cysteine has three potential partners in the fully reduced tetramer, but only one in the half-oxidized structure) (Reproduced from Cristian, L. et al. *Proc. Natl. Acad. Sci. USA.* (2003) 100, 14772–14777. "Copyright (2003) National Academy of Sciences, U.S.A.")

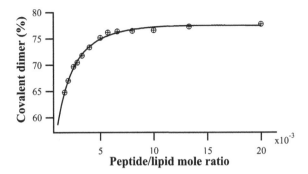

Fig. 6 Thiol-disulfide exchange measurements of the M2 peptide incorporated in DLPC bilayers. Calculated percent of covalent dimer as a function of peptide/DLPC ratio. The reduced form of M2TM peptide was incorporated into vesicles at various peptide/phospholipid (DLPC) mole ratios. The reversibility of the disulfide cross-linking process in lipid bilayers was assessed by repeating the equilibration reactions, starting with the oxidized dimer rather than the reduced form of the peptide. Irrespective of whether the initial starting material was oxidized or reduced, the same equilibrium distribution was observed. The solid line is the best fit to the data as described in the text (Reproduced from Cristian, L. et al. *Proc. Natl. Acad. Sci. USA.* (2003) 100, 14772–14777. "Copyright (2003) National Academy of Sciences, U.S.A.")

the amount of monomer and disulfide-bonded dimer (as a result of tetramerization) at equilibrium for various peptide-to-lipid ratios (Fig. 6).

The dissociation constants and equilibria were fit to the experimental data shown in Fig. 6 using Scheme 2, which illustrates the coupling between disulfide formation and tetramerization [3]. The free energy of tetramer dissociation (K_1) was then calculated by $\Delta G = -RT \ln K_1$.

Thus, the method can be applied to investigate the association energetics of oligomeric systems of higher order than dimers.

4.2.2 The Effect of Lipid Composition on TM Helix Association

This method can also be used to evaluate the effect of lipid composition on the association of TM peptides. We have applied this method to study how various lipids (of different chain length) affect the association M2TM into tetramers. This work revealed that association of the M2 TM helices is modulated by phospholipid acyl chain length and cholesterol presence (Figs. 7 and 8). In particular, the association was found to be stronger in POPC than in DLPC and became stronger upon cholesterol addition. These results are in line with previous reports of self-association of peptides as an adjustment response to variations in the bilayer thickness [22, 23]. We thus demonstrate that the thiol-disulfide method can provide information how lipid chain length and bilayer thickness modulate peptide interactions.

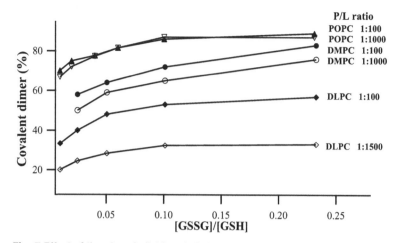

Fig. 7 Effect of the phospholipid acyl chain length on the covalent dimer formation. The peptide–phospholipid mole ratios (*P/L*) for each lipid investigated were as indicated. DMPC and DLPC show dependence on the *P/L* ratio, whereas POPC shows no dependence (Reproduced from Cristian, L. et al. *Proc. Natl. Acad. Sci. USA.* (2003) 100, 14772–14777. "Copyright (2003) National Academy of Sciences, U.S.A.")

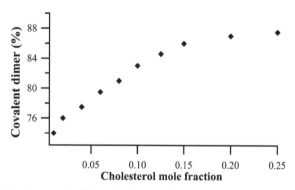

Fig. 8 Effect of increasing the mole fraction of cholesterol on the extent of disulfide cross-linking. The peptide–phospholipid mole ratio was 1:500, and the thiol-disulfide exchange equilibrations were performed at a GSSG/GSH ratio of 0.25 (Reproduced from Cristian, L. et al. *Proc. Natl. Acad. Sci. USA.* (2003) 100, 14772–14777. "Copyright (2003) National Academy of Sciences, U.S.A.")

4.2.3 The Topology of TM Helix Association in Lipid Bilayers

Another application of this method is to experimentally determine the orientation of the alpha-helical peptides in membrane-like environments such as lipid bilayers. As an example, we cite a synthetic peptide designed based on the serine zipper motif, found in various crystal structures of membrane proteins and which can adopt a parallel or antiparallel conformation [24]. To understand what features are required to define one topology versus the other, a serine zipper peptide (SerZip) was designed and the thiol-disulfide method was utilized to determine which topology was preferred [11].

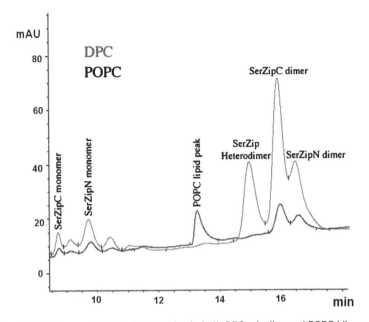

Fig. 9 HPLC traces of the SerZip peptides in both DPC micelles and POPC bilayers. Mixtures of reduced SerZipN and SerZipC peptides incorporated in DPC micelles (*gray trace*) and POPC bilayers (*black trace*) were incubated in a glutathione redox buffer (75/25, GSSG/GSH mol ratio), each allowed to equilibrate, and then analyzed by HPLC. Figure shows the disulfide-bonded SerZipN, SerZipC homodimers, the disulfide-bonded SerZipN–SerZipC heterodimer, as well as SerZipN and SerZipC monomers as identified by MALDI-TOF. Their identities were also confirmed by individual HPLC runs of the disulfide-bonded SerZipN and SerZipC homodimers. The *large peak* present in the SerZip-POPC chromatogram (in *black*) corresponds to the POPC lipid. The different intensities observed for the two homodimers are the result of using unequal amounts of the thiol-free peptides in the topology assay (Reproduced from [11] with permission from Elsevier)

A flexible Gly linker with terminal cysteine residue was added at either the C or N termini of the SerZip peptide to probe association in lipid bilayers using reversible cysteine cross-linking.

The ability of the SerZip peptides to form parallel versus antiparallel dimers was assessed using a similar protocol as the one described in Subheading 4.1.2 and which is schematically illustrated in Fig. 5. A mixture of N- and C-terminally cysteine-labeled SerZip peptides that can form both parallel and antiparallel dimers was incubated in a glutathione redox buffer (at different GSSG/GSH mole ratios) that allows the peptide monomers and disulfide bonds to rearrange. Redox equilibrium experiments were initially conducted in detergent micelles (DPC). In DPC micelles, a mixture of parallel and antiparallel dimers was observed, with a slight preference of the peptides for parallel homodimers, as indicated by the experimentally obtained molar ratio of 1:1:1 of the two homodimers and heterodimer, versus the 1:1:2 ratio expected for a random association of the peptide chains (Fig. 9).

The preference of the peptides to form disulfides in phospholipid bilayers of different chain length, POPC ($C_{16:0}C_{18:1}$PC) and DLPC ($C_{12:0}$ PC), was also examined. Only homodimers were formed in both POPC and DLPC, indicating a very strong preference for the parallel dimer, results which contrast with the observation of a mixture of parallel and antiparallel dimers in micelles (Fig. 9). Thus, the topology assumed by the peptides is very sensitive to the environment and bilayers provide conformational specificity for a parallel versus an antiparallel orientation and the thiol-disulfide method can distinguish between these orientational preferences.

5   Concluding Remarks

The experimental method established herein should provide a broadly applicable tool for thermodynamic studies of membrane protein folding, particularly oligomerization and protein–protein interactions. The thiol-disulfide exchange method presents a few advantages over the other methods used in membrane helix association studies. For example, AUC has been extensively used to determine the oligomerization state and association energetics of TM helices [5, 9, 24] and a modified protocol has been also employed to determine the topology of TM helices [8]. However, the main disadvantage of the AUC experiments in measuring TM helix association free energies is that the method is limited to membrane proteins solubilized in detergent micelles and is not useful for experiments in lipid vesicles [5, 6]. In contrast, the thiol-disulfide exchange method can be applied in both detergent micelles and lipid vesicles environments. In addition, the application of AUC in quantifying the TM helix association energetics is limited to the protein-to-detergent molar ratio range used in the experiments. For some TM helices, AUC cannot provide quantitative association energetics [8]. This problem can be efficiently solved by the thiol-disulfide exchange method which can be applied to a wide range of protein-to-detergent molar ratios [3, 8]. Another method widely used in both detergent and lipid systems is FRET [12, 25]. Both FRET and thiol-disulfide methods require the introduction of a label. FRET requires introduction of donor and acceptor fluorophores and requires additional control experiments to ensure that the labels do not influence the studied systems. The thiol-disulfide exchange requires incorporation of cysteine if a natural cysteine is not available. The cysteine should be extended with a flexible linker such as Gly3. Considering that two cysteine-Gly3 linkers could extend up to ~20 Å, while interhelical distances in dimers vary by only ~0–4 Å, the flexibility of the linker should easily accommodate any subtle differences in helix packing as long as the helices pack in a parallel manner. In this way, the calculated association energetics and topology only depend on the properties of the TM helices and are independent of the cysteine labeling.

While the thiol-disulfide exchange method is an efficient tool to measure TM helix association or topology of oligomeric helical bundles when the oligomeric state is pre-identified (particularly self-associating, homo-oligomeric systems), the method is difficult to apply to large, multi-pass membrane proteins. Despite the caveats, this method serves as an efficient tool, together with other existing methods, for characterizing TM helix–helix interactions in membrane-mimicking environments.

6 Notes

1. Degas all solutions used in the experiments (main buffer, glutathione buffers).

2. Determine the concentration of the oxidized and reduced glutathione in each stock solution by UV measurements, using an absorption coefficient of 273 M^{-1} cm^{-1} at 280 nm in the case of oxidized glutathione and by titration with Ellman's reagent (DTNB), measuring the absorbance at 412 nm and using an absorption coefficient of 13,600 M^{-1} cm^{-1} in the case of reduced glutathione [7].

3. Conduct equilibrations of the samples under anaerobic conditions in airtight cap screwed vials to prevent adventitious oxidation.

References

1. You M, Li E, Hristova K (2005) FRET in liposomes: measurements of TM helix dimerization in the native bilayer environment. Anal Biochem 340:154–164

2. Duong MT, Jaszewski TM, MacKenzie KR (2007) Changes in apparent free energy of helix-helix dimerization in a biological membrane due to point mutations. J Mol Biol 371:422–434

3. Cristian L, Lear JD, DeGrado WF (2003) Use of thiol-disulfide equilibria to measure the energetics of assembly of transmembrane helices in phospholipid bilayers. Proc Natl Acad Sci USA 100:14772–14777

4. Chen L, Novicky L, Hristova K (2010) Measuring the energetic of membrane protein dimerization in mammalian membranes. J Am Chem Soc 132:3628–3635

5. DeGrado WF, Gratkowski H, Lear JD (2003) How do helix-helix interactions help determine the folds of membrane proteins? Perspectives from the study of homo-oligomeric helical bundles. Protein Sci 12:647–665

6. MacKenzie KR, Fleming KG (2008) Association energetics of membrane spanning α-helices. Curr Opin Struct Biol 18:412–419

7. Regan L, Rockwell A, Wasserman Z, DeGrado WF (1994) Disulfide crosslinks to probe the structure and flexibility of a designed four-helix bundle protein. Protein Sci 3: 2419–2427

8. Zhang Y, Kulp DW, Lear JD et al (2009) Experimental and computational evaluation of forces directing the association of transmembrane helices. J Am Chem Soc 131: 11341–11343

9. Cristian L, Lear JD, Degrado WF (2003) Determination of membrane protein stability via thermodynamic coupling of folding to thiol-disulfide interchange. Protein Sci 12: 1732–1740

10. Stouffer AL, Ma C, Cristian L et al (2008) The interplay of functional tuning, drug resistance,

and thermodynamic stability in the evolution of the M2 proton channel from the influenza A virus. Structure 16:1067–1076

11. North B, Cristian L, Fu Stowell X et al (2006) Characterization of a membrane protein folding motif, the Ser zipper, using designed peptides. J Mol Biol 359:930–939

12. Choma C, Gratkowski H, Lear JD, DeGrado WF (2000) Asparagine-mediated self-association of a model transmembrane helix. Nat Struct Biol 7:161–166

13. Walshaw J, Woolfson DN (2001) Socket: a program for identifying and analysing coiled-coil motifs within protein structures. J Mol Biol 307:1427–1450

14. Mason JM, Arndt KM (2004) Coiled coil domains: stability, specificity, and biological implications. Chembiochem 5:170–176

15. Yu YB (2002) Coiled-coils: stability, specificity, and drug delivery potential. Adv Drug Deliv Rev 54:1113–1129

16. Acharya A, Rishi V, Vinson C (2006) Stability of 100 homo and heterotypic coiled-coil a-a′ pairs for ten amino acids (A, L, I, V, N, K, S, T, E, and R). Biochemistry 45:11324–11332

17. Wagschal K, Tripet B, Lavigne P, Mant C et al (1999) The role of position a in determining the stability and oligomerization state of alpha-helical coiled coils: 20 amino acid stability coefficients in the hydrophobic core of proteins. Protein Sci 8:2312–2329

18. Lamb RA, Zebedee SL, Richardson CD (1985) Influenza virus M2 protein is an integral membrane protein expressed on the infected-cell surface. Cell 40:627–633

19. Holsinger LJ, Lamb RA (1991) Influenza virus M2 integral membrane protein is a homotetramer stabilized by formation of disulfide bonds. Virology 183:32–43

20. Pinto LH, Holsinger LJ, Lamb RA (1992) Influenza virus M2 protein has ion channel activity. Cell 69:517–528

21. Wang C, Takeuchi K, Pinto LH, Lamb RA (1993) Ion channel activity of influenza A virus M2 protein: characterization of the amantadine block. J Virol 67:5585–5594

22. Ren J, Lew S, Wang J, London E (1999) Control of the transmembrane orientation and interhelical interactions within membranes by hydrophobic helix length. Biochemistry 38: 5905–5912

23. Killian JA (1998) Hydrophobic mismatch between proteins and lipids in membranes. Biochim Biophys Acta 1376:401–416

24. Adamian L, Liang J (2002) Interhelical hydrogen bonds and spatial motifs in membrane proteins: polar clamps and serine zippers. Proteins: Struct Funct Genet 47: 209–218

25. Yin H, Slusky JS, Berger BW et al (2007) Computational design of peptides that target transmembrane helices. Science 315: 1817–1822

Chapter 2

Measurement of Transmembrane Peptide Interactions in Liposomes Using Förster Resonance Energy Transfer (FRET)

Ambalika Khadria and Alessandro Senes

Abstract

Present day understanding of the thermodynamic properties of integral membrane proteins (IMPs) lags behind that of water-soluble proteins due to difficulties in mimicking the physiological environment of the IMPs in order to obtain a reversible folded system. Despite such challenges faced in studying these systems, significant progress has been made in the study of the oligomerization of single span transmembrane helices. One of the primary methods available to characterize these systems is based on Förster resonance energy transfer (FRET). FRET is a widely used spectroscopic tool that provides proximity data that can be fitted to obtain the energetics of a system. Here we discuss various technical aspects related to the application of FRET to study transmembrane peptide oligomerization in liposomes. The analysis is based on FRET efficiency relative to the concentration of the peptides in the bilayer (peptide:lipid ratio). Some important parameters that will be discussed include labeling efficiency, sample homogeneity, and equilibration. Furthermore, data analysis has to be performed keeping in mind random colocalization of donors and acceptors in liposome vesicles.

Key words Integral membrane proteins, Transmembrane helix, Energetics, Thermodynamic equilibrium, Free energy of association, FRET in liposomes

1 Introduction

Free energy measurements of transmembrane (TM) helix association in single and multispan membrane proteins are important to understand the mechanisms behind vital cellular processes such as membrane protein folding and signal processing [1]. However, research in the field of integral membrane proteins has considerably lagged behind in comparison to that of soluble proteins [2]. This is due to the difficulty in obtaining experimental systems that match the nature of their complex native environment, the bilayer, and conditions in which reversible association/unfolding can be established. Moreover, since the unfolded state of membrane proteins retains considerable amount of helical component [1] (in

Giovanna Ghirlanda and Alessandro Senes (eds.), *Membrane Proteins: Folding, Association, and Design,*
Methods in Molecular Biology, vol. 1063, DOI 10.1007/978-1-62703-583-5_2, © Springer Science+Business Media, LLC 2013

opposition to soluble proteins where the unfolded state largely lacks a secondary structure), monitoring unfolding is a significant challenge.

To overcome these challenges, a strong focus has been applied toward a more approachable folding question—the oligomerization of single-pass TM domains—a question that retains the core process that is important for the folding of membrane proteins, the association of the transmembrane helices. Several biophysical tools have been developed to measure TM helix associations in a number of environments, from detergent micelles to lipid vesicles and even biological membranes. Analytical ultracentrifugation (AUC), Förster resonance energy transfer (FRET), and thiol-disulfide exchange [2–4] are biophysical methods that are suitable for measuring the association of TM peptides in artificial environments. Genetic methods based on the conditional expression of a reporter gene such as TOXCAT [5] and GALLEX [6] are useful for measuring homo- and hetero-association of TM helices in the natural inner membrane of *Escherichia coli*.

In this chapter we focus on FRET-based studies of TM association in artificial liposomes. FRET involves excitation of the ground electronic state of a donor molecule followed by non-radiative transfer of energy from the excited state of the donor to an appropriate acceptor. Since FRET occurs only when two suitable fluorophores (a donor and acceptor molecule) are located within ~10 nm of each other, it can be used as a measure of molecular proximity [1, 4, 7]. The two fluorophores must have a significant spectral overlap between the emission spectrum of the donor and the excitation spectrum of the acceptor, a sufficiently high quantum yield, and a favorable dipole–dipole orientation (Fig. 1). When the fluorophores come in close proximity, energy is transferred non-radiatively, resulting in quenching of donor fluorescence and increase (sensitization) in acceptor fluorescence.

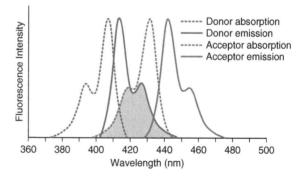

Fig. 1 Spectral overlap between a FRET pair. *Blue curves* depict the donor excitation and emission spectra and *green curves* depict the acceptor excitation and emission spectra. Note significant overlap (*shaded region*) between the donor emission and the acceptor excitation, leading to a good spectral overlap, one of the requirements for a FRET pair

In the case of TM helix–helix interactions, each peptide is labeled with either a donor or an acceptor molecule, and they are solubilized in artificial hydrophobic environments like detergents and lipids. Because TM peptides are insoluble in water and only experience the "hydrophobic volume" of the solution, their concentration is expressed as mole fraction, i.e., the number of moles of peptide relative to number of moles of the "true solvent" (detergent or lipid), yielding a peptide:detergent or peptide:lipid ratio. In addition to mole fractions, the second parameter that is often varied in a typical FRET experiment is the relative percentage of the donor- and acceptor-labeled peptides, which can help in determining the specific oligomeric state (i.e., dimer, trimer) of the system being studied [8]. Once a donor/acceptor system has been equilibrated, the FRET efficiency can be calculated by monitoring the degree of donor quenching in the presence of acceptor (as discussed in this chapter) or by monitoring the increase in acceptor emission. Spectral overlap between the donor and acceptor also causes contamination of the FRET signal due to Acceptor Spectral Bleedthrough, which refers to direct excitation of the acceptor by radiation at the donor excitation wavelength (Fig. 2). This must be subtracted from the acceptor emission intensity in the FRET samples while calculating FRET efficiency using acceptor sensitization (not used in this chapter).

FRET has been shown to yield association energetics in TM peptides in detergents [9] as well as lipids [3, 7]. It can also be used for studying hetero-oligomers in which cases the calculations are applied taking into account the various possible equilibria present in the system. This chapter discusses the experimental procedure to

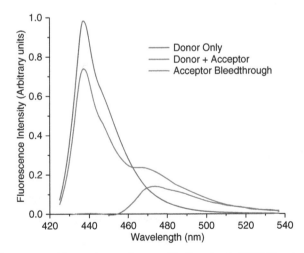

Fig. 2 Donor quenching and acceptor sensitization due to FRET. The decrease in the donor emission intensity is used to calculate the FRET efficiency. The sensitization of acceptor emission (FRET) is evident from the increased fluorescence intensity compared to direct excitation (acceptor bleedthrough) of the acceptor at the donor excitation wavelength

measure helix–helix *self* association in lipid vesicles and the relationship between mole fraction and FRET efficiency to calculate the association of affinity. An outline of the sample preparation techniques, including solid-phase peptide synthesis, labeling technique, HPLC purification, and MS characterization, is presented. This is followed by a detailed description of the FRET experimental layout and data collection. In addition, there is a discussion of possible challenges faced in data interpretation due to false positives and false negatives that can arise from adventitious interactions in vesicles [1, 7, 10], inefficient labeling efficiency of the molecules [10], or light scattering.

2   Materials

2.1   Solid-Phase Peptide Synthesis

1. Automated peptide synthesizer.
2. Amino acids.
3. Resin: Fmoc-PAL-PEG.
4. Activator: 2-(1H-7-Azabenzotriazol-1-yl)-1,1,3,3-tetramethyl uronium hexafluorophosphate methanaminium (HATU).
5. Base: *N*-methylmorpholine (NMM).
6. Deprotecting agent: 20 % Piperidine.
7. Solvent: Dimethylformamide (DMF), Dichloromethane (DCM), 1-methyl-2-pyrrolidone (NMP).

2.2   N-Terminal Labeling with Fluorophores

1. Fluorophore for solid-phase coupling.
2. Solvent: DMF, NMP.
3. Base: *N,N*-Diisopropylethylamine (DIPEA).
4. Activator: Benzotriazol-1-yl-oxytripyrrolidinophosphonium hexafluorophosphate (PyBOP) or (7-Azabenzotriazol-1-yloxy)tripyrrolidinophosphonium hexafluorophosphate (PyAOP).

2.3   HPLC Purification and Mass Spectrometry (MS)

1. Reversed phase semi-preparative and analytical column.
2. HPLC instrument.
3. Solvents: Acetonitrile, Isopropanol, water, Trifluoroacetic acid (TFA).
4. Lyophilizer maintained below –80 °C.
5. Glassware: glass screw cap vials, pear-shaped glass flask.
6. Compressed N_2 gas.
7. Mass spectrometry facility.

2.4 Solubilization of TM Peptides in Lipids and FRET Measurements

1. Solvents: Hexafluoroisopropanol (HFIP), Chloroform, Trifluoroethanol (TFE).

2. 1-Palmitoyl-2-oleoyl-*sn*-glycero-3-phosphocholine (POPC) lipid (Avanti Polar Lipids).

3. Purified TM peptides.

4. Compressed N_2 gas.

5. Liposome buffer (10 mM Sodium Phosphate Buffer, 500 mM NaCl, pH 7.0).

6. Dry ice, acetone, 37 °C water bath.

7. TFE lined glass screw cap vials, glass syringes.

8. UV/Vis spectrophotometer, quartz cuvettes.

9. Plate reader or Fluorimeter (If using a plate reader, the plate reader should contain a monochromator, such as the Tecan M1000 which provides for a Fluorescence intensity scan instead of point measurements from fixed emission wavelength filters).

3   Methods

3.1 Fmoc Solid-Phase Peptide Synthesis

1. Review the amino acid sequence of the peptide/peptides of interest, adding positively charged Lys residues to increase aqueous solubility of the highly hydrophobic sequence (*see* **Note 1**) [11].

2. Decide the scale of synthesis and the resin to be used (*see* **Note 2**). Weigh the calculated amount of resin, use the appropriate solvent (DMF/NMP) and activator (HATU), and set up the automated synthesis. A review of solid-phase peptide synthesis should be consulted for standard protocols and optimizations (*see* **Note 2**) [12].

3. Take the resin containing the completed peptide, check for correct sequence by MS (*see* **Note 3**), and then proceed with on-resin N-terminal labeling of the peptide with the fluorescent dye (*see* **Note 4**).

   This step onward, perform all experiments protected from light, till the end of the chapter.

3.2 N-Terminal Labeling with Fluorescent Dye

1. Divide the resin with peptide in half and proceed with two separate manual N-terminal couplings using donor and acceptor fluorescent dyes comprising a good FRET pair. If the dye is not carboxylic acid derivatized to enable Fmoc chemistry on a deprotected amine group, then an appropriate linker is attached to the amine group first (e.g., aminohexanoic acid, mini-PEG), followed by attachment of the fluorescent dye.

2. Use DMF and NMP as solvents and PyBOP/PyAOP as activators using the protocol optimized for hydrophobic sequences [13] (*see* **Note 5**).

3. Cleave the labeled peptide, precipitate using cold ether, and dry it under a stream of compressed N_2 gas with agitation by a glass rod to avoid clumping.

4. Blanket the peptide with a stream of N_2 or Ar, seal the cap with parafilm, and store it at –20 °C.

3.3 Reversed Phase HPLC Purification

1. Solubilize a small amount of the peptide in the appropriate solvent for HPLC purification (*see* **Note 6**). Blanket the rest of the peptide with a stream of N_2 or Ar, seal the cap with Teflon, and store it in –20 °C.

2. Filter the sample with a 0.22 μm filter.

3. Using the appropriate column for use, inject the sample and follow a gradient of 2–100 % Buffer B in 98 min (1 % per minute). Monitor the chromatogram at 280 nm if Trp or Tyr is present and at the wavelength of the fluorescent dye (*see* **Note 7**).

4. Collect the various fractions and analyze them using MS to identify the peak of interest (*see* **Note 7**).

5. Optimize the gradient to achieve better separation and collection of individual peaks.

6. Perform multiple HPLC runs using the same method and pool in the desired fractions from the various runs in a glass pear-shaped flask of appropriate volume such that the total volume doesn't exceed a third of the flask.

7. Flash-freeze the sample in the flask, rotating it over liquid N_2 with a tilt to increase the surface:volume ratio for better lyophilization. Attach it to the lyophilizer, cover it with foil, and leave it overnight till it forms a dry powder.

8. Take the flask out of the lyophilizer and empty its contents into a pre-weighed glass vial with screw cap. Weigh the vial with the sample in it and note the weight of the final sample.

3.4 Solubilization of Peptides and Lipids to Make Stock Solutions

1. Dissolve 1 mg of powdered POPC lipid (Avanti Polar lipids) in 1 mL of 1:1:1 HFIP:Chloroform:TFE in a screw cap glass vial and keep it tightly sealed. Calculate the number of moles/μL according to Scheme 1.

2. Turn on the spectrophotometer and set up the method for protein concentration scan including the absorbance wavelength of Trp (280 nm) and that of the fluorescent dyes.

3. Perform a blank scan with 1:1:1 HFIP: Chloroform: TFE (solvent).

$$Molecular\ weight = 760.10g/mol$$

$$No.\ of\ moles = weight/molecular\ weight$$

$$= \frac{1mg}{760.1mg/mol}$$

$$= 1.31\ umol\ in\ 1\ ml,\ or\ \underline{1.31\ nmol/\ \mu L}$$

Scheme 1 Calculation for number of moles for 1 mg/mL POPC lipid stock

Concentration of the fluorescent label on peptide:

$$\mathbf{[Fluorophore]} = A_{max}/\varepsilon_{Fluorophore}l$$

where A_{max} is absorbance at λ_{max} of the dye
$\varepsilon_{Fluorophore}$ is extinction coefficient of the dye at that absorbance
l is path length of the cuvette (usually 1cm)

Concentration of the peptide based on Trp absorbance (λ=280nm)

$$\mathbf{[Peptide]} = A_{280} - (A_{max} * CF)/\varepsilon\ l$$

where A_{280} is absorbance at 280nm
CF is the correction factor that adjusts for absorbance at 280 nm by the fluorophore, and is given by
$$A_{280} / A_{max}$$
ε is extinction coefficient of the peptide (calculated based on no. of Trp, Tyr and Cys)

After calculating the concentration in terms of Molarity, calculate the number of moles in the stock solutions of the peptides similar to Scheme 1.

Calculating the labeling efficiency

$$\mathbf{Percent\ labeling = [Fluorophore]/[Peptide] * 100}$$

Note: For peptides that do not have a Trp or Tyr to get accurate concentration from UV/VIS Spectrophotometry, CD can also be used to determine peptide concentration(3)

Scheme 2 Calculation of protein concentration and labeling efficiency

4. Scoop out a tiny amount of donor-labeled peptide and add it to 100 μL of the solvent. Label this as 'stock 1'.

5. Make a 1:10 dilution of stock 1 (call it 'stock 2') and add it to the appropriate cuvette for protein concentration scan (*see* **Note 8**).

6. Calculate concentration of the peptides and the labeling efficiency according to Scheme 2 (*see* also Subheading 3.8).

7. Repeat **steps 4–6** for the acceptor-labeled peptide.

3.5 Setup of the Peptide:Lipid Ratios for FRET

The goal of this experimental setup is to achieve a large concentration range of peptide or a range of peptide:lipid ratios which spans the spectrum from a "no FRET" sample to a "maximum FRET" sample. Since liposomes, unlike detergent micelles, do not "communicate" with other liposomes to exchange peptide molecules, titrating the solution with more lipids will not dilute the peptide in the lipid solvent, or change the effective mole fraction of the peptides in the liposomes. Thus to obtain sufficient number of data points for an accurate curve fitting to obtain the energetics of the system, samples spanning a wide range of peptide:lipid ratios have to be prepared in separate tubes. Lipids can be equilibrated to form multilamellar vesicles (MLVs) as described later. It has been shown that FRET efficiencies for helix–helix interactions in large unilamellar vesicles (LUVs) are comparable to that of MLVs, and thus this protocol follows setting up FRET interactions in MLVs [3].

1. Once the peptide stocks for donor- and acceptor-labeled peptides and the lipid stocks are ready, start labeling small 12×35 mm screw cap vials for the FRET experiments.

2. Cover the labeling on the tubes with a tape to prevent the labels from being washed away by acetone in the dry-ice bath used in future steps.

3. From the peptide stock solutions, calculate the volume required for 50 pmol of the donor- and acceptor-labeled peptides (*see* **Note 9**).

4. Take a corning black 96-well plate, and add the required volume for 50 pmol of the donor peptide, say 8.9 µL, into one well. If the well volume is 75 µL, add $75 - 8.9 = 66.1$ µL of liposome buffer into it and mix by pipetting up and down a few times. In another well, do the same for the acceptor peptide. In a third well, add 75 µL of liposome buffer only.

5. Take the plate to the plate reader and run two separate fluorescence scans—one spanning the absorbance spectrum of the donor and another of the acceptor (*see* **Note 10**).

6. For the "donor-only" well, set the excitation maximum wavelength of the donor such that there is maximum overlap between the donor emission and acceptor excitation spectrum. If the emission spectrum looks noisy, try changing the parameters of the instrument (gain, PMT voltage, slit width, etc.) or increase the amount of peptide to get a smooth signal. The goal is to achieve a high enough concentration of the peptide for good signal to noise ratio (*see* **Note 11**).

7. Again, for the "acceptor-only" well, set the same excitation wavelength used in **step 6**. This step is to test for spectral bleedthrough between the donor and acceptor pairs. A minimal emission peak at the acceptor emission maxima on being excited by the donor excitation maxima is a sign of a good FRET pair.

8. Perform both **steps 5** and **6** for the "liposome buffer" well. This sample serves as a blank for data analysis.

9. At this point, start making the peptide:lipid ratio calculations. An example of the volumes is given in Table 1, based on the example in Subheading 3.4, **step 4**.

10. Pipette these solutions using glass syringes into the small glass screw cap vials arranged in increasing peptide:lipid ratio on a vial rack and vortex vigorously to mix them (Fig. 3a).

11. Take each vial in a fume hood and slowly evaporate the solvent using a light stream of compressed N_2 gas till all the solvent has evaporated, leaving a thin whitish lipid film at the bottom/ sides of the vial (Fig. 3b) (*see* **Note 12**).

12. Don't screw the caps on the vials. Cover the tray with vials with aluminum foil to protect it from light, and pierce holes on the foil at the mouth of each vial.

Table 1
Sample 15 with the highest amount of lipid and same amount of unlabeled peptides serves as a scattering control to subtract any background scattering from the lipids alone

Sample No.	Mole fraction (peptide:lipid ratio)	Volume of donor stock (µL)	Volume of acceptor stock (µL)	Volume of 1 mg/mL POPC lipid stock (µL)
1	1:100	Calculate for 50 pmol (8.9 µL)	Calculate for 50 pmol	Calculate for 10 nmol (=7.7)
2	1:200	Ditto	Ditto	20 nmol = 15.4
3	1:500	Ditto	Ditto	50 nmol = 38.5
4	1:1,000	Ditto	Ditto	100 nmol = 77
5	1:2,000	Ditto	Ditto	200 nmol = 154
6	1:5,000	Ditto	Ditto	500 nmol = 385
7	1:10,000	Ditto	Ditto	1 mmol = 770
8	Control 1	Ditto	x	10 nmol = 7.7
9	Control 2	Ditto	x	20 nmol = 15.4
10	Control 3	Ditto	x	50 nmol = 38.5
11	Control 4	Ditto	x	100 nmol = 77
12	Control 5	Ditto	x	200 nmol = 154
13	Control 6	Ditto	x	500 nmol = 385
14	Control 7	Ditto	x	1 mmol = 770
15	No fluorophore control (contains 100 pmol unlabeled peptide)	x	x	1 mmol = 770
16	No liposome control	Ditto	Ditto	x

Sample 16 serves as a control for any FRET arising out of peptide aggregates in solution in the case of incomplete incorporation of peptides into vesicles

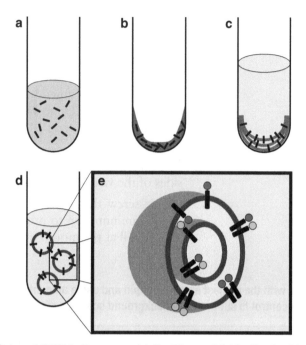

Fig. 3 Setup of FRET in liposomes. (**a**) Peptides and lipids dissolved in solvent and mixed together in a glass vial. (**b**) The solvent is evaporated using a stream of Nitrogen gas to leave behind a thin lipid film containing the peptide molecules. (**c**) The lipid film is hydrated using aqueous buffer, vortexed, and freeze–thawed to achieve proper equilibration and (**d**) formation of multilamellar vesicles containing the peptides. (**e**) The various pairs that will be contributing to FRET efficiency due to association of peptides and just colocalization, as well as "dark pairs" formed by donor–donor, acceptor–acceptor associations

13. Place the covered tray inside a vacuum dessicator overnight (*see* **Note 13**).

14. The next day, add 75 μL (or 100 μL depending on the volume of the wells in the plate or the cuvette of the fluorimeter) of liposome buffer into each vial, screw its cap on tightly, and vortex vigorously for about a minute (Fig. 3c). The samples with higher lipid will turn more turbid. Note the extent of turbidity in these samples.

3.6 Preparation of Multilamellar Lipid Vesicles (MLVs) Using Freeze–Thaw Cycles

1. Prepare a dry-ice/acetone bath and a water bath set at 37 °C.

2. Place the tray with vials on the dry-ice/acetone bath for about 1 min. Make sure all the samples are frozen and the labels are not getting washed off.

3. Now place the tray on the water bath, shaking them mildly. A crackling sound will signify thawing of the liquids. Continue thawing till the sound subsides.

4. Repeat **steps 2–3** alternately for 2–3 cycles.

5. Now note the turbidity of the samples again. Decrease in turbidity signifies formation of MLVs (*see* **Note 14**). At this point, the donor and acceptor peptides should have equilibrated with the lipid and with each other based on their affinity and the peptide-to-lipid ratio (Fig. 3d).

6. Pipette out the samples from the vials into a 96-well corning plate according to Table 1. Do not pipette up and down to avoid formation of air bubbles in the sample as they can lead to light-scattering effects.

3.7 Measurement of FRET and Data Analysis

1. Take the plate to the plate reader, insert the plate, and adjust the settings for the plate (or alternatively use a fluorimeter).

2. Set the excitation wavelength as that for maximum spectral overlap for the pair and collect the fluorescence emission scans spanning the emission wavelengths of both the donor and acceptor fluorophores. Maintain the same scan for each sample.

3. The "Control" samples in Table 1 serve as the "no FRET" controls for their corresponding sample row. For instance, sample 1 in Table 1 has a 1:100 peptide-to-lipid ratio where there are 50 pmol of donor and acceptor each. Control 1 therefore has 50 pmol of the donor only, with the same amount of lipid as sample 1.

4. Calculate percent FRET, $E(\%)$, by the wavelength of the emission maximum of the donor in the absence and presence of the acceptor according to Scheme 3.

3.8 Discussion

A lot of factors need to be taken into account for accurate FRET intensity measurements. One of the most important factors is the labeling efficiency of peptides. Hydrophobic peptides pose a greater level of difficulty for their labeling and purification, and alternative methods for labeling have been discussed [13]. Once this step has been optimized, the next level of difficulty arises in accurate quantification of these peptides. An unlabeled peptide is commonly quantified using UV/Vis spectrophotometry where the Trp and Tyr absorbance at 280 nm is measured and protein concentration determined using Beer's law. In the case of a labeled peptide, the absorbance of the peptide sample at 280 nm comprises three components:

1. Trp/Try absorbance of labeled peptide

2. Trp/Tyr absorbance of unlabeled peptide

3. Absorbance of the dye at 280 nm

Thus it is important to differentiate the individual components for accurate quantification. This is done using the "Correction Factor of the dye" and the labeling efficiency of the peptide

$$E_{observed} = [(I_D - I_{DA})/ I_D] \qquad ...Eq. (1)$$

where I_D nd I_{DA} are the donor emission maximum intensities of samples containing only donor-

labeled proteins (controls)and samples with both donor-and acceptor-labeled proteins, respectively.

$$E_{expected} = \frac{[D][A]}{[D][D]+[D][U]+[D][A]}$$

$$\text{where } [D] = [D]^{total} * L^D$$
$$[A] = [A]^{total} * L^A$$
$$\text{and } [U] = [D]^{total}(1-L^D)+[A]^{total}(1-L^A)$$

([D] is concentration of labeled donor-peptide, L_D is labeling efficiency of donor-peptide and
[D] ^{total} is total donor-peptide, [A] is concentration of labeled acceptor-peptide, L_A is labeling
efficiency of donor-peptide and [A] ^{total} is total acceptor-peptide, and [U] is concentration of unlabeled
peptide)

Fraction Dimer $= E_{observed} / E_{expected}$

$= $ **Dimer / Total Peptide**

$$= 2X_{dimer} / (2X_{dimer} + X_{monomer}) \qquad ...Eq. (2)$$

where X is the mole fraction of the peptide.

Finally, mole fraction concentrations can be used to calculate a partition coefficient by

$$K_x = [X_{dimer}]/ [X_{monomer}]^2 \ ...Eq. (3)$$

which is an equilibrium constant, and the free energy of association is calculated by

$$\Delta G_x = -RT \ln (K_x)$$

Scheme 3 Calculation of percent FRET efficiency

(Scheme 2). However, to be able to use the calculations in Scheme 2, another important parameter needs to be established, the molar extinction coefficient, $\varepsilon_{Fluorophore}$ *in the given solvent system.* Vendors selling dyes usually provide the molar extinction coefficient values of dyes at their A_{max} in an aqueous buffer at a particular pH. But these values of $\varepsilon_{Fluorophore}$ and A_{max} significantly change in different solvents (Fig. 4). For accurate determination of equimolar ratios of the peptides, it is very important to characterize the behavior of the fluorophores in the solvent system being used. This can be done by plotting a calibration curve of the A_{max} of the dye in the given solvent versus concentration. Sometimes it is difficult to dissolve a known amount of dye in the organic solvent at a measurable concentration. In that case it is advisable to make the stock solution in a buffer that it has been characterized in, and then use the provided molar extinction coefficient and λ_{max} values to calculate the stock concentration. One should make this concentration high enough such that a very small volume can be used to dilute into the organic solvent to obtain the calibration curve. Then serial

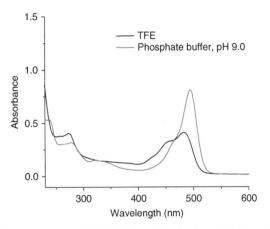

Fig. 4 Absorbance spectra of FITC in Phosphate Buffer, pH 9.0, and in TFE. The solvent has a blue shift effect on the λ_{max} and its molar extinction coefficient ε

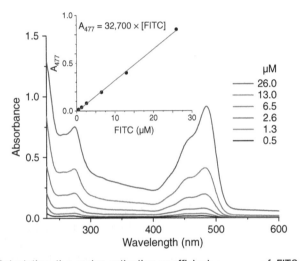

Fig. 5 Calculating the molar extinction coefficient $\varepsilon_{Fluorophore}$ of FITC in TFE. Absorbance scans of increasing concentrations of FITC dye in organic solvent. A_{max} (477 nm) is plotted versus concentration and the slope of the curve yields the new molar extinction coefficient $\varepsilon_{Fluorophore}$ of FITC (32,700 M^{-1} cm^{-1}) in TFE (*inset*). The ratio of A_{280} to A_{max} yields the new correction factor for FITC in TFE required for calculating the degree of labeling. Note that the wavelength of maximum absorbance shifts and that the molar extinction coefficient at A_{max} decreases more than twofold compared to that in Phosphate Buffer pH 9.0

dilutions should be made to obtain samples of known concentration in organic solvent, absorbance scans should be taken (Fig. 5), and A_{max} of the dye in that solvent versus concentration should be plotted (Fig. 5 inset). The slope of the curve will provide for the $\varepsilon_{Fluorophore}$ at the new λ_{max} of the dye in the organic solvent. Once the behavior of the dye has been characterized in the given solvent in this manner, the concentration and labeling efficiency of the peptide can be calculated according to Scheme 2 and utilized in Scheme 3 to calculate the FRET efficiency values.

Apart from labeling efficiency, another contributor towards the observed FRET intensity is random colocalization due to proximity of donors and acceptors, which needs to be subtracted from the steady state FRET observed. Depending on the level of quantification necessary for the experiment at hand, there are several ways by which this can be taken into account [14, 15]. The peptide pairs that will be forming in a self-associating system of TM peptides will be the random proximity pairs, along with DA, DD, AA, DU, AU, and UU (where D is donor-labeled peptide, A is acceptor-labeled peptide, and U is unlabeled peptide). Figure 3e shows a schematic representation of the possible pairs. Finally, once all these parameters have been characterized, the FRET efficiency can be calculated according to Scheme 3. Contributions of proximity and sequence-dependent association to FRET efficiency can be distinguished by spiking the existing donor–acceptor-labeled samples with unlabeled peptide at the time of preparation. If the FRET efficiency is primarily due to sequence-dependent association, then addition of unlabeled peptides will lead to a decrease in the FRET efficiency [3, 7].

For a homodimer, the FRET efficiency, a measure of association, (Scheme 3, Eq. (2)) is calculated according to Scheme 3, Eq. (1), and the data is fitted to a curve using Scheme 3, Eq. (3) to yield the association constant and the free energy values [3, 7].

4   Notes

1. Lysine tags increase solubility and decrease peptide aggregation in hydrophobic sequences by providing for unfavorable electrostatic interactions [11]. The number of Lys residues incorporated and their positioning varies depending on the TM peptide sequence, and a detailed description of this optimization can be found in [11]. Addition of extra amino acids will also provide for a flexible linker between the peptide and the dye with which it will be labeled. It is important, however, to make sure that the linker length keeps the fluorophores within the limit of their Förster radius in all orientations.

2. It is advisable to begin with a smaller scale of synthesis as hydrophobic peptides usually need some optimization of the protocol for synthesis. A 25 μmol scale is a good starting point and is enough for preliminary experiments. Choice of resin is important—a low load resin (e.g., Fmoc-PAL-PEG resin from Applied Biosystems, with a 0.18 mmol/g loading capacity) will decrease aggregation potential of the peptides on the resin. After an assessment of the hydrophobicity of the final sequence, a "brute force" method can be applied for the preliminary trial.

For instance, in a Val-Val-Thr-Ala-His sequence, the difficult couplings are Val-Val, Val-Thr, and Thr-Ala. For these couplings, one could use double or triple coupling and extended coupling times, whereas for the Ala-His coupling, a single coupling should suffice. This approach has been utilized before [16] and is found to be effective for particularly hydrophobic sequences.

3. CHCA (4-Chloro-α-Cyanocinnamic Acid) is one of the best universal matrices for performing MALDI-TOF in peptides [17]. Typically a barely visible amount of peptide is dissolved in about 100 μL of Acetonitrile in water (percentage depending on solubility of peptide). The matrix can be prepared as follows: 10 mg of CHCA matrix in 1 mL of 50:50 Acetonitrile: water with 0.1 % TFA. Then various ratios of the matrix:peptide solutions are made and 1–2 μL of the final samples are spotted on the target for MALDI-TOF analysis.

4. For FRET, the peptide has to be labeled with a donor or acceptor molecule. In this procedure, the labeling will be done on the N-terminus manually, on-resin, using commercially available carboxylated fluorescent dyes. An important point is to program the automated synthesis to terminate the last coupling step before deprotection. This will keep the final N-terminal residue Fmoc protected and prevent unwanted reactions till the peptide is ready for labeling. Also, it is advisable that the peptide on-resin remains solvated till the next step to reduce aggregation. Before proceeding with the labeling process, it is better to take out a tiny amount of the resin (a touch with the spatula), deprotect it and cleave it using standard procedures [12], and then analyze it using MS. If the peptide has been synthesized correctly, deprotect the last amino acid on the entire resin manually and proceed with the labeling.

5. For quantitative FRET experiments, it is important to obtain high labeling yields of peptides, as the separation of the labeled from the unlabeled species in the case of hydrophobic peptides is difficult by HPLC [3, 7, 18–21]. Standard on-resin N-terminal labeling yields have been found to be very low for hydrophobic peptides, and thus a method for higher labeling efficiency of hydrophobic peptides was developed, utilizing larger amounts of dyes and coupling agents [13].

6. Dissolve minimum amount of peptide (~0.5 mg) in minimum amount of DMF. If the peptide does not readily dissolve, try other solvents like HFIP, THF, and TFE. Then slowly add water dropwise, till the peptide just starts precipitating, turning slightly murky. It is important to have a high enough percentage of water in the solution (≥ 60 %) for the peptide to bind the reversed phase column, but also maintain solubility of the peptide at the same time. If using TFA for solubilization, do not exceed the final TFA concentration to be above 5 %.

7. The Reversed Phase HPLC system usually consists of an *n*-alkylsilica-based sorbent from which the solutes are eluted. The most commonly used *n*-alkyl ligand is C-18, but for hydrophobic peptides, C2, C4, C8, phenyl, and cyanopropyl ligands provide better separation [22]. It is good practice to first perform a blank run with 100 % Buffer A till a stable baseline is obtained and then a blank run with the solvent the sample is dissolved in, before injecting the sample. Choice of columns for hydrophobic peptides varies and there are various parameters that can be changed for optimization. Usually a small volume (~10 µL) of the sample is injected into an analytical column first and analyzed for the retention time of the sample elution with a linear gradient from 2 % to 100 % Buffer B (Acetonitrile with 0.1 % TFA) over 30 min. For hydrophobic peptides which do not elute easily, sometimes using 60 % Isopropanol in Acetonitrile with 0.1 % TFA can be used for Buffer B.

If LC-MS facilities are available, it is better to perform a run on an analytical column and monitor the masses of the peaks as the elute. The conditions can then be transferred to a semi-prep column having the same *n*-alkyl silica group for fraction collection.

8. Sometimes it is necessary to make higher dilutions of the initial stock solution of the peptide for accurate concentration determination. In that case, make serial dilutions till you get an absorbance value between 0 and 1, use the most diluted stock for concentration determination, and calculate the concentrations of the original stock by scaling up. Later in the experiment this stock will be used for solubilization with lipid and then involve evaporation of the solvents, and thus it is easier to have a higher stock concentration to minimize the volume of solvent to be evaporated.

9. If the stock concentration of the peptide is 0.2 mg/mL and you have 500 µL of that solution, the concentration in terms of number of moles is 5.6 pmol/µL. Thus, to use 50 pmol of peptide, use $50/5.6 = 8.9$ µL of the stock solution. Calculate the stock concentration of each donor- and acceptor-labeled peptide in number of moles per microliter and start with a total of 100 pmol of 1:1 peptide (donor/acceptor) to begin the peptide/lipid molar ratios.

10. A fluorimeter can also be used in place of a plate reader, and the sample volumes in that case have to be according to the volume of the fluorimeter cuvette (~10–20 µL higher) than the cuvette volume to prevent the beam from hitting the meniscus of the solution leading to scattering. It has been shown that volume of the sample does not change the FRET efficiency as long as the amounts of lipid and peptide are the same [3].

11. It should be kept in mind that increasing the peptide amount will also mean increasing the amount of lipid to maintain a peptide:lipid ratio. If the lipid concentration is too high, light scattering by the lipid will interfere with the FRET results, so a balance has to be achieved. For a true scattering control, use the same amount of unlabeled peptide in the highest amount of lipid being used in the experiment to check for any scattering.

12. It may take ~20 min for vials with higher volumes of solvent to evaporate. The higher the lipid content, the longer will it take to evaporate and the more visible will the lipid film be.

13. This step can also be substituted by adding the samples in a lyophilizer for 2–3 h if the experiment has to be finished on the same day.

14. Freeze–thaw cycles are generally required for equilibration of buffers, salts, etc. MLVs, prepared by premixing the proteins and lipids in organic solvent and hydrating the mixture, are in equilibrium after a single freeze–thaw cycle [3]. When thin lipid films are hydrated, stacks of liquid crystalline bilayers become fluid and swell as shown in Fig. 3c. The hydrated lipid sheets detach during agitation and self-close to form MLVs preventing interaction of the hydrocarbon core of the bilayer from interacting with water at the edges. It has been shown that the presence of unlabeled peptide helps in equilibration of the peptides in the lipids and leads to a faster decrease in turbidity upon just one freeze–thaw cycle.

References

1. Popot JL, Engelman DM (1990) Membrane protein folding and oligomerization: the two-stage model. Biochemistry 29:4031–4037
2. MacKenzie KR, Fleming KG (2008) Association energetics of membrane spanning alpha-helices. Curr Opin Struct Biol 18:412–419
3. You M, Li E, Wimley WC et al (2005) Förster resonance energy transfer in liposomes: measurements of transmembrane helix dimerization in the native bilayer environment. Anal Biochem 340:154–164
4. DeGrado WF, Gratkowski H, Lear JD (2003) How do helix-helix interactions help determine the folds of membrane proteins? Perspectives from the study of homo-oligomeric helical bundles. Protein Sci 12:647–665
5. Russ WP, Engelman DM (1999) TOXCAT: a measure of transmembrane helix association in a biological membrane. Proc Natl Acad Sci 96:863–868
6. Schneider D, Engelman DM (2003) GALLEX, a measurement of heterologous association of transmembrane helices in a biological membrane. J Biol Chem 278:3105–3111
7. Li E, You M, Hristova K (2005) Sodium dodecyl sulfate–polyacrylamide gel electrophoresis and Förster resonance energy transfer suggest weak interactions between fibroblast growth factor receptor 3 (FGFR3) transmembrane domains in the absence of extracellular domains and ligands. Biochemistry 44:352–360
8. Li M, Reddy LG, Bennett R et al (1999) A fluorescence energy transfer method for analyzing protein oligomeric structure: application to phospholamban. Biophys J 76:2587–2599
9. Fisher LE, Engelman DM, Sturgis JN (1999) Detergents modulate dimerization, but not helicity, of the glycophorin A transmembrane domain. J Mol Biol 293:639–651
10. Yevgen MM, Posokhov O (2008) A simple "proximity" correction for Förster resonance energy transfer efficiency determination in membranes using lifetime measurements. Anal Biochem 380(1):134–136

11. Rath A, Deber AM (2013) Design of transmembrane peptides: coping with sticky situations. Meth Mol Biol 1063:197–210

12. Amblard M, Fehrentz J-A, Martinez J et al (2005) Fundamentals of modern peptide synthesis. Methods Mol Biol 298:3–24

13. Stahl PJ, Cruz JC, Li Y et al (2012) On-the-resin N-terminal modification of long synthetic peptides. Anal Biochem 424:137–139

14. Wolber PK, Hudson BS (1979) An analytic solution to the Förster energy transfer problem in two dimensions. Biophys J 28:197–210

15. Wimley WC, White SH (2000) Determining the membrane topology of peptides by fluorescence quenching. Biochemistry 39:161–170

16. Fisher LE, Engelman DM (2001) High-yield synthesis and purification of an α-helical transmembrane domain. Anal Biochem 293:102–108

17. Leszyk JD (2010) Evaluation of the new MALDI matrix 4-chloro-α-cyanocinnamic acid. J Biomol Tech 21:81–91

18. Schick S, Chen L, Li E et al (2010) Assembly of the M2 tetramer is strongly modulated by lipid chain length. Biophys J 99:1810–1817

19. Chen L, Merzlyakov M, Cohen T et al (2009) Energetics of ErbB1 transmembrane domain dimerization in lipid bilayers. Biophys J 96:4622–4630

20. Merzlyakov M, Hristova K (2008) Forster resonance energy transfer measurements of transmembrane helix dimerization energetics. Methods Enzymol 450:107–127

21. You M, Spangler J, Li E et al (2007) Effect of pathogenic cysteine mutations on FGFR3 transmembrane domain dimerization in detergents and lipid bilayers. Biochemistry 46:11039–11046

22. Aguilar M-I (2004) Reversed-phase high-performance liquid chromatography, HPLC of peptides and proteins. Humana Press, New Jersey, 251:9–22

Chapter 3

Measuring Transmembrane Helix Interaction Strengths in Lipid Bilayers Using Steric Trapping

Heedeok Hong, Yu-Chu Chang, and James U. Bowie

Abstract

We have developed a method to measure strong transmembrane (TM) helix interaction affinities in lipid bilayers that are difficult to measure by traditional dilution methods. The method, called steric trapping, couples dissociation of biotinylated TM helices to a competitive binding by monovalent streptavidin (mSA), so that dissociation is driven by the affinity of mSA for biotin and mSA concentration. By adjusting the binding affinity of mSA through mutation, the method can obtain dissociation constants of TM helix dimers ($K_{d,dimer}$) over a range of six orders of magnitudes. The $K_{d,dimer}$ limit of measurable target interaction is extended 3–4 orders of magnitude lower than possible by dilution methods. Thus, steric trapping opens up new opportunities to study the folding and assembly of α-helical membrane proteins in lipid bilayer environments. Here we provide detailed methods for applying steric trapping to a TM helix dimer.

Key words Membrane protein folding, Steric trap, Glycophorin A dimer, Transmembrane helix interaction, Monovalent streptavidin, Biotinylation

---

1 Introduction

Lateral interactions between transmembrane (TM) helices mediate the folding of α-helical membrane proteins, receptor assembly, and signaling in cell membranes [1–3]. To understand the physico-chemical principles of these processes, significant efforts have been made to develop the methods for measuring thermodynamic stability of TM helix interactions in lipid environments [4–9]. The methods developed so far such as analytical centrifugation, reversible thiol-disulfide exchange, and Förster energy transfer (FRET), primarily rely on dilution of model TM peptides in detergents or lipids to a concentration range in which associated and dissociated populations become comparable. However, these methods cannot be applied to strongly interacting TM helices because there are practical limitations on how far the interacting helices can be diluted in lipid bilayers.

Giovanna Ghirlanda and Alessandro Senes (eds.), *Membrane Proteins: Folding, Association, and Design*,
Methods in Molecular Biology, vol. 1063, DOI 10.1007/978-1-62703-583-5_3, © Springer Science+Business Media, LLC 2013

To overcome limitations on dilution, we developed a new method called steric trapping that involves coupling TM helix dissociation to a competitive binding event [10, 11]. In general, the basic tool kit includes an affinity tag which is covalently attached to the target TM helix; a bulky tag-binding protein which competes with the association, sterically trapping the dissociated monomers; and a method to sensitively monitor the dissociation of the target TM helix interaction or the binding of the tag-binding protein.

So far we have used a biotin affinity tag and monovalent streptavidin (mSA) as the tag binder. The highly specific biotin-monovalent streptavidin [12] binding pair is a versatile tool that can be conveniently embedded into the steric trapping strategy. The biotin tag can be efficiently attached to the target protein by enzymatic or chemical reactions. The tag-binding protein mSA can be produced in large quantities, and the exceptionally strong biotin affinity ($K_{d,biotin} \sim 10^{-14}$ M) can be adjusted to a desirable range by mutations to achieve reversible conditions for thermodynamic measurements [11, 13]. We used the glycophorin A transmembrane domain (GpATM) that forms a stable dimer in lipid environments as well-characterized model system [14, 15]. Here, we describe the practical steps for measuring the thermodynamic stability of strong TM helix interactions in lipid bilayers using the steric trap method.

2   Principle and Strategy

A biotin acceptor peptide [16] (BAP) is inserted N-terminal to the GpATM domain fused to staphylococcal nuclease [14] (SN-GpATM) (Fig. 1a). We attach a biotin tag on each BAP using biotin ligase BirA. The two biotin tags must be located sufficiently close to each other and also to the target interaction site (GpATM) so that steric overlap would occur if two mSA molecules were bound in the dimer. Thus, the first biotin can be bound in the dimer with the intrinsic affinity of the biotin–mSA pair, but a second mSA can only bind if the dimer dissociates to monomers. Thus, the affinity of the second mSA binding is coupled to dissociation and therefore reflects dimer affinity (Fig. 1b). We therefore observe biphasic binding. At low concentrations, we observe an mSA-binding phase reflecting the intrinsic biotin affinity, but at higher concentrations we observe a second phase reflecting the weaker binding that is coupled to dissociation. Binding is detected by monitoring the increase of fluorescence of a pyrene label on SN-BAP-GpATM. As the reasonable concentration range of SN-BAP-GpATM is limited, it is necessary to tune the affinity of the mSA to a practical level. For example, if the affinity of mSA is too high relative to the free energy of dissociation, we can't detect a second phase. If the affinity of mSA is too low, we have to raise the mSA concentrations to impractically high levels to see

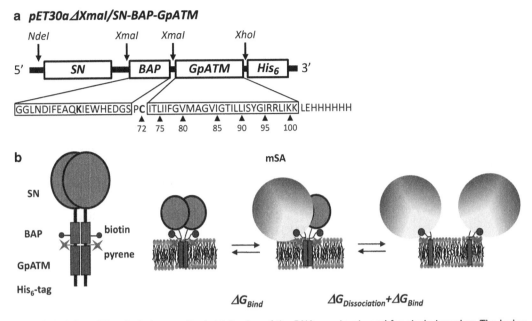

a *pET30aΔXmaI/SN-BAP-GpATM*

Fig. 1 Principles of the steric trap method. (**a**) Design of the DNA construct used for steric trapping. The lysine residue in the biotin acceptor peptide (BAP, *bold*) can be biotinylated by biotin ligase BirA. A unique cysteine residue (*bold*) at the position 72 was introduced for pyrene labeling. (**b**) (Changed to grey scale) Reaction scheme of the steric trap method. The second monovalent streptavidin (mSA)-binding affinity is modulated by the stability of the dimer due to the coupling of dimer dissociation and mSA binding: $\Delta G_{dissociation} + \Delta G_{bind}$. Reprinted with permission from Hong and Bowie [26]. Copyright 2011 American Chemical Society

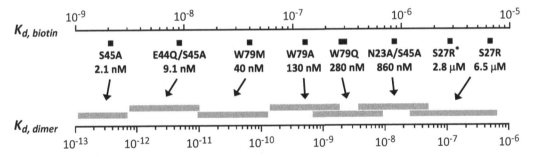

Fig. 2 Library of mSA mutants with various intrinsic biotin-binding affinities (*top*, $K_{d,biotin}$). If the affinity of mSA is too high, binding is insensitive to the contribution from dissociation. If the affinity is too low, impractically high concentrations of mSA are required to drive dissociation. Therefore, depending on the stability of the dimer ($K_{d,dimer}$), a mutant mSA with an optimal intrinsic biotin affinity should be selected (*bottom*). The dissociation constants shown reflect binding to monomeric SN-BAP-GpATM and must be determined for each system. mSA-S27R (*asterisk*) has a stronger biotin-binding affinity in the negatively charged membranes than in neutral lipid environments. Modified and reprinted with permission from Hong and Bowie [26]. Copyright 2011 American Chemical Society

the second binding phase. We therefore generated a series of mSA mutants whose biotin affinity ($K_{d,biotin}$) ranges from $\sim 10^{-9}$ to $\sim 10^{-6}$ M (Fig. 2) [11]. The biotin affinity of mSA must be tuned to the stability of the dimer interaction of interest. The method can measure strong TM helix interactions because dissociation can be thermodynamically driven by using the high affinity of mSA and by the concentration of mSA.

3   Materials

3.1   Description of DNA Vector Encoding SN-BAP-GpATM

The *SN-GpATM* gene was inserted into a kanamycin-resistant pET30a vector using *NdeI* and *XhoI* restriction sites to fuse the His$_6$-tag to the C-terminus of SN-GpATM (Fig. 1a). The intrinsic *XmaI* site in the pET30a vector was eliminated by site-directed mutagenesis to generate pET30aΔ*XmaI*. A DNA cassette encoding the biotin acceptor peptide (BAP; amino acid sequence: GGLNDIFEAQKIEWHEDGSP; DNA sequence: 5′-phospho-CC GGG GGC CTG AAC GAC ATC TTC GAG GCT CAG AAA ATC GAA TGG CAC GAG GAC GGC TGT C, underlined nucleotides designates the complementary sequence ligated to the *XmaI* cut-site) was then inserted into the single *XmaI* restriction site, which was located right before the 5′-end of *GpATM* [14]. A unique Cys residue was introduced at Gly72 by site-directed mutagenesis for thiol-specific labeling with a fluorophore.

3.2   Expression and Purification of SN-BAP-GpATM

The procedure is modified from the reported protocol [14, 17]. SN-BAP-GpATM with the C-terminal His$_6$-tag is expressed into inclusion bodies, extracted by decyl maltoside (DM) detergents, and purified by affinity chromatography using Ni-NTA resin.

1. Transform *E. coli* BL21(DE3)RP cells (Agilent) with pET30aΔ*XmaI/SN-BAP-GpATM*. Pick a colony and grow in 5 ml Luria-Bertani media (LB: 10 g tryptone, 5 g yeast extract, 10 g NaCl per liter) containing 0.05 g/l kanamycin for 6 h. Then inoculate 1 l terrific broth (TB: 12 g tryptone, 24 g yeast extract, 4 ml glycerol, 23.12 g KH$_2$PO$_4$, 125.41 g K$_2$HPO$_4$ per liter) containing 0.05 g/l kanamycin at a 1–400 dilution.

2. Grow at 30 °C overnight (~16 h). The next day, increase the temperature to 37 °C and when the OD$_{600nm}$ reaches 1.5–2.0, induce the expression of SN-BAP-GpATM with 1 mM IPTG. Incubate for additional 3 h.

3. Harvest and resuspend the cells in 30 ml of 50 mM Tris–HCl, 5 mM EDTA buffer solution (pH 8.0) containing 1 mM PMSF (Thermo Scientific) and 1 mg/ml hen egg lysozyme (Sigma). Freeze and thaw the resuspended cells three times in dry ice/ethanol and a 42 °C water bath, respectively. Add CaCl$_2$ to a final concentration of 10 mM to activate the nuclease domain of SN-BAP-GpATM. Digest the chromosomal DNAs for 20 min at 4 °C.

4. Centrifuge the lysate at 31,000 × g for 15 min in 45Ti rotor. Resuspend the pellets in 40 ml of 50 mM Tris–HCl, 1 M NaCl

buffer (pH 8.5) containing 2 % (w/v) decylmaltoside (DM) (Anatrace, Sol-grade), using a tissue homogenizer to solubilize SN-BAP-GpATM.

5. Centrifuge the resuspension at 57,000 × g for 1 h in a 45Ti rotor. Collect the supernatant (approximately 40 ml) and add imidazole stock solution (4 M, pH 8.0) to a final concentration of 10 mM. Mix with 4 ml of Ni-NTA resin.

6. After stirring gently for 2 h at 4 °C, pack the slurry into to a column (Econo-Pac chromatography column, Bio-rad). Wash the protein-bound resin with 10 resin volumes of 0.5 % DM, 50 mM Tris–HCl, 200 mM NaCl, 10 mM imidazole buffer solution (pH 8.5) and elute with 10-resin volumes of the same buffer containing 200 mM imidazole.

3.3 In Vitro Biotinylation of SN-BAP-GpA

1. Biotinylate 40 ml of 50–100 μM SN-BAP-GpATM in 50 mM Tris–HCl, 200 mM NaCl, 200 mM imidazole buffer (pH 8.5) containing 5 mM ATP (Sigma), 1 mM D-biotin (Sigma), 5 mM magnesium acetate, 2 mM TCEP (Tris(2-carboxyethyl) phosphine hydrochloride, Pierce), and 2 μM BirA overnight at room temperature with gentle stirring [18].

2. After the removal of aggregates by centrifugation, concentrate the sample to less than 3 ml using Amicon Ultra centrifugal filters (MWCO = 30 kDa, Millipore).

3. Equilibrate an Econo-Pac 10DG desalting column (Bio-rad) with 0.5 % DM, 50 mM Tris–HCl, and 200 mM NaCl buffer solution (pH 8.5).

4. Remove excess biotin and imidazole using the desalting column. Estimate the SN-BAP-GpATM concentration using UV absorbance at 280 nm ($\varepsilon_{molar} = 22{,}920$ M^{-1} cm^{-1}). The typical yield of biotinylated SN-BAP-GpATM is 500–1,000 μM in 4 ml total volume from 1 l of TB culture.

5. Biotinylation efficiency can be quickly tested using SDS-PAGE by looking for the disappearance of the SN-BAP-GpA band after mixing with an excess of wild-type mSA. The mSA–biotin interaction remains intact in the 2 % SDS sample buffer (Fig. 3).

6. Store the sample frozen at –20 °C.

3.4 Thiol-Specific Pyrene Labeling of SN-BAP-GpA

1. Reduce 5 ml of 100 μM of SN-BAP-GpATM-G72C in 2 % DM, 50 mM Tris–HCl, 200 mM NaCl (pH 8.5) with ten times molar excess of TCEP for 90 min at room temperature.

2. Add 20 times molar excess of thiol-reactive N-(1-pyrenemethyl) iodoacetamide (Invitrogen) from a 20 mg/ml stock solution dissolved in DMSO to the SN-BAP-GpATM solution. Incubate the reaction in the dark with gentle stirring at room temperature for ~16 h.

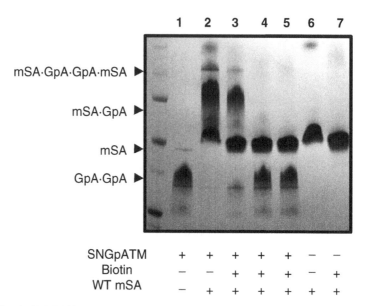

Fig. 3 SDS-PAGE showing the binding of mSA to the biotinylated SN-BAP-GpATM-G72C. The unique Cys is blocked by a thiol-reactive pyrene label (*see* Subheading 3.4) to prevent the formation of an intermolecular disulfide bond. *Lane 1*: 10 μM of SN-BAP-GpATM in 20 mM DM; *lane 2*: 40 μM of WT mSA was added; *lane 3*: 40 μM of WT mSA was added first and incubated for 1 day with excess biotin; *lane 4*: 40 μM of WT mSA was added first and incubated for 7 days with excess biotin; *lane 5*: 40 μM of wild-type mSA pre-incubated with 2 mM biotin was added to 10 μM of SN-BAP-GpATM. All samples were not heated. Reproduced from Hong et al. [11]

3. To remove unreacted pyrene label, immobilize the SN-BAP-GpATM on Ni-NTA resin and extensively wash with 1 % DM, 5 mM imidazole, 50 mM Tris–HCl, 200 mM NaCl buffer (pH 8.5). Elute with the same buffer containing 200 mM imidazole. Concentrate the eluate to less than 3 ml using Amicon Ultra centrifugal filters (MWCO = 30 kDa, Millipore).

4. Remove the excess imidazole using a Bio-rad 10DG desalting column and concentrate the eluate to less than 1 ml. Determine the pyrene-labeling efficiency from the total protein concentration obtained by Bio-rad Dc assay and the absorbance of pyrene at 345 nm (ε_{molar} (280 nm) = 22,920 M^{-1} cm^{-1}, ProtParam tool from ExPASy Proteomics Server). The efficiency typically ranges from 70 to 90 %. The total yield of the fluorescently labeled SN-BAP-GpATM is 200–600 μM in 500–1,000 μl.

3.5 Expression of Active and Inactive Streptavidin

The expression and purification procedure of streptavidin is modified from the reported protocols [12, 19]. The inclusion bodies of active subunit with the C-terminus His$_6$-tag and inactive subunit (triple mutant N23A/S27D/S45A) without a tag are separately expressed, multiply washed with Triton X100, and solubilized by GdnHCl.

1. Transform *E. coli* BL21(DE3)RP cells with the pET21a vector encoding streptavidin (active or inactive) [12]. Grow a colony in LB media containing 0.1 g/l ampicillin at 37 °C overnight.

2. Inoculate the culture in a larger volume of LB media at 1–20 dilution in the presence of 0.1 g/l ampicillin. Induce the expression of streptavidin at $OD_{600nm} = 0.7$ by adding 1 mM IPTG and continue incubating for 3 h.

3. Resuspend the harvested cells in 1/100 culture volume of 50 mM Tris–HCl, 0.75 M sucrose, 1 mg/ml hen egg lysozyme (pH 8.0) followed by 3 cycles of freeze/thaw in a dry ice/ethanol and a 42 °C water bath, respectively.

4. Treat the lysate with 1 mg/ml bovine pancreas DNase (Sigma) and 10 mM $CaCl_2$ for 1 h at 4 °C and centrifuge at $31,000 \times g$ for 15 min at 4 °C using a 45Ti rotor.

5. Wash the pellet by resuspension in 1/150 culture volume of 50 mM Tris–HCl (pH 8.0), 1.5 M NaCl, 0.5 % TritonX100 (Sigma) using a tissue homogenizer. Centrifuge the resuspension at $45,000 \times g$ for 15 min at 4 °C using a 45Ti rotor.

6. Repeat the washing step two more times.

7. Further wash the resulting pellet with the same buffer without TritonX100 and centrifuge again. Solubilize the precipitated inclusion body by 1/250 culture volume of 6 M GdnHCl (pH 2.0) with the aid of a tissue homogenizer. Centrifuge the sample at 27,000 rpm for 45 min using a 45Ti rotor.

8. Measure the OD_{280nm} of the supernatant. Typical OD_{280nm}'s for the final solutions were 20–40 (from 2 to 4 l culture) for active and 60–80 for inactive variant (from 8 to 12 l culture).

3.6 Refolding and Purification of Monovalent Streptavidin (mSA)

The original protocol of Howarth et al. [12] is employed with minor modifications. It involves the reassembly of tetrameric streptavidin from an unfolded mixture of His_6-tagged active subunits and untagged inactive subunits. The singly tagged mSA is then isolated by the selective elution from a Ni-NTA resin at low imidazole concentrations (Fig. 4).

1. Mix the GdnHCl-solubilized stocks of His_6-tagged active streptavidin and the untagged inactive streptavidin in a 1:4 molar ratio.

2. Add this mixture quickly drop by drop to a chilled and vigorously stirred 20 mM sodium phosphate, 200 mM NaCl buffer solution (pH 7.5) in Erlenmeyer flask to 1–40 final dilution.

3. Incubate the solution at 4 °C for 15 min and centrifuge at $9,000 \times g$ for 30 min at 4 °C in a GS-3 rotor. Mix the supernatant

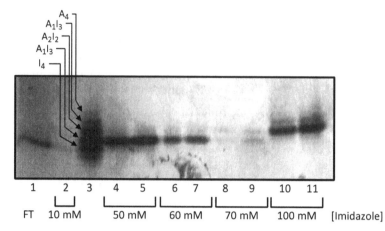

Fig. 4 Purification of mSA by Ni-NTA affinity chromatography. The major fraction of mSA is selectively eluted at 50 and 60 mM imidazole (*lanes 4–7*). *Lane 1*: FT (flow-through), *lane 2*: wash, *lane 3*: molecular weight standard for the mixture of inactive (I_4), monovalent (A_1I_3), divalent (A_2I_2), trivalent (A_3I_1), and tetravalent (A_4) streptavidin prepared by refolding of 1:1 mixture of unfolded active and inactive subunits

with 1/100 volume of washed/equilibrated Ni-NTA resin and stir gently for 1 h at 4 °C.

4. Collect the protein-bound Ni-NTA resin by centrifugation in a bench-top centrifuge and then pack into a Bio-rad disposable chromatography column (4 ml resin/column).

5. Wash the column with 10 resin volumes of wash buffer (10 mM imidazole, 20 mM sodium phosphate, 200 mM NaCl , pH 7.5). Elute the mSA with 20 resin volume of 50 mM imidazole buffer and then by 10 resin volume of 60 mM imidazole buffer.

6. Confirm the presence of mSA by Tris-Gly 12 % SDS-PAGE without sample heating (Fig. 4). During the run, cool the gel box in ice to minimize heat-induced subunit rearrangements.

7. Concentrate the collected 50 mM and 60 mM imidazole eluates to less than 40 ml using 30 kDa cutoff Amicon ultrafiltration filters and dialyze against 20 mM sodium phosphate, 200 mM NaCl, pH 7.5 to remove the excess imidazole (Spectra/Por 10 kDa cutoff membranes).

8. To maximize the purity of mSA, mix the dialyzed protein solution again with 3 ml of washed/equilibrated Ni-NTA resin after centrifugation. Repeat the purification in the Ni-NTA column. Concentrate the resulting eluted proteins and dialyze into the same buffer without imidazole as above.

9. The final yield of mSA varied significantly depending on the mutations on the active subunit. For 1 l of a total refolding solution, we typically obtained 0.5–2 ml of a 150–250 μM

mSA *tetramer* using $\varepsilon_{molar}(280$ nm$) = 167,760$ M^{-1} cm^{-1} (ProtParam tool, ExPASy Proteomics Server).

4 Methods

4.1 Determination of Intrinsic Biotin Affinity of mSA to SN-BAP-GpA

Measurements of the intrinsic biotin affinities of the mSA mutants are essential for obtaining the thermodynamic dissociation constant of the SN-BAP-GpATM dimer (*see* Eq. 3). To measure the intrinsic affinity, we employed the mutant G83I of GpATM, which exists predominantly in a monomeric form when highly diluted, so that the monomer–dimer equilibrium does not contribute to mSA binding [20]. The intrinsic biotin affinities ($K_{d,biotin}$) of weaker biotin binding mSA mutants (S27R, N23A/S45A, and W79A [19]) were determined by direct binding measurements between pyrene-labeled G83I SN-BAP-GpATM mutant and mSA (*see* **step 1** below, Figs. 2 and 5a). The biotin affinities of the stronger binding mutants were obtained by competition for binding between a mutant with known biotin affinity and a mutant with unknown affinity. We developed a FRET-based competition assay for this purpose (*see* **steps 2–5** below and Fig. 5b).

1. Titrate 300 nM pyrene-labeled G83I in 100 mM DM with mSA-S27R and -N23A/S45A mutants, respectively. Monitor the pyrene-fluorescence changes at 390 nm upon excitation at 330 nm and fit the data to the Eq. 1, which yielded $K_{d,biotin} = 6.5 \pm 2.0$ µM and $K_{d,biotin} = 870 \pm 210$ nM, respectively. For mSA-W79A, perform a binding measurement with 30 nM pyrene-labeled G83I SN-BAP-GpATM in 100 mM DM, from which $K_{d,biotin} = 130 \pm 30$ nM was obtained. Prepare for all samples in 20 mM phosphate, 200 mM NaCl, pH 7.5 buffer solution (Fig. 5a).

2. Measure $K_{d,biotin}$'s of stronger biotin binding mSA's (S45A [21], E44Q/S45A, W79M, W79Q) by a FRET-based competition assay because we cannot readily detect the labeled SN-BAP-GpATM mutant fluorescence at the dilutions required for straightforward binding assays (Fig. 5b). Label the G83I SN-BAP-GpATM mutant with Oregon Green488 iodoacetamide (Invitrogen) at Cys72 using the same protocol as the pyrene-labeling described above. Label mSA-S45A and -W79A mutants with amine-reactive Qsy7 succinimidyl ester (Invitrogen). Typically, 4–6 Lys residues were labeled per tetrameric mSA. In this experimental design, Oregon Green488 acts as a fluorescent donor and Qsy7 as a nonfluorescent acceptor (dark quencher).

3. First, pre-incubate 400 nM Oregon Green488-labeled G83I SN-BAP-GpATM mutant with Qsy7-labeled mSA-W79A in 100 mM DM (20 mM sodium phosphate, 200 mM NaCl,

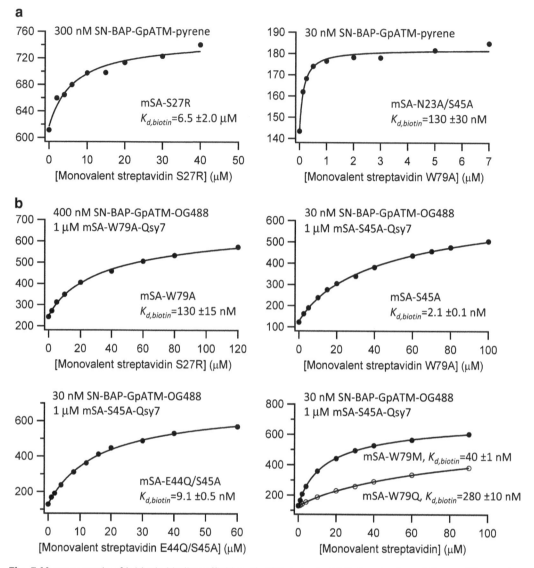

Fig. 5 Measurements of intrinsic binding affinities of mSA mutants. (**a**) Determination of $K_{d,biotin}$ of the weaker biotin binding mutant mSAs by direct binding measurement with the highly diluted pyrene-labeled G83I SN-BAP-GpATM mutant in DM micellar solution. (**b**) Determination of $K_{d,biotin}$ of the stronger biotin binding mutant mSA using FRET-based competition assays. Reproduced from Hong et al. [11]

pH 7.5). Titrate this complex with unlabeled mSA-S27R with a known $K_{d,biotin} = 6.5$ µM and incubated overnight at room temperature. Monitor the dequenched fluorescence from Oregon Green488 at 530 nm with an excitation wavelength of 485 nm and fit the data to Eq. 2 [21], which yielded $K_{d,biotin} = 130 \pm 15$ nM (Fig. 5b, upper left). This $K_{d,biotin}$ agreed well with the result obtained from the direct binding assay, which confirmed the validity of this method.

4. Determine $K_{d,biotin}$ of mSA-S45A by titrating the pre-incubated 1 μM Qsy7-labeled mSA/400 nM of Oregon Green488-labeled G83I SN-BAP-GpATM complex with unlabeled mSA-W79A ($K_{d,biotin} = 130 \pm 15$ nM), which yielded $K_{d,biotin}$(mSA-S45A) $= 2.1 \pm 0.1$ nM (Fig. 5b, upper right).

5. For mSA-E44Q/S45A (Fig. 5b, lower left), -W79M, and -W79Q mutants (Fig. 5b, lower right), pre-incubate 400 nM of G83I SN-BAP-GpATM with an appropriate concentration of mSA-S45A ($K_{d,biotin} = 2.1 \pm 0.1$ nM; 4 μM for mSA-E44Q/S45A titrations, 1 μM for mSA-W79M and -W79Q titrations), mix with increasing concentrations of the corresponding mutant mSA, and incubate overnight at room temperature. The fitting of the dequenching curves yielded $K_{d,biotin} = 9.1 \pm 0.5$ nM (mSA-E44Q/S45A), 40.0 ± 1.4 nM (mSA-W79M), and 280 ± 14 nM (mSA-W79Q), respectively.

$$\theta_{bound} = A_1 \frac{\left(P_o + [mSA] + K_{d,biotin}\right) - \left(\left(P_o + [mSA] + K_{d,biotin}\right)^2 - 4P_o[mSA]\right)^{1/2}}{2P_o} + A_2 \quad (1)$$

P_o: Total SN-BAP-GpATM concentration (fixed); A_1, A_2: Amplitude of binding phase and initial fluorescence level, respectively (fitted); [mSA]: Total mSA concentration (variable); $K_{d,biotin}$: Dissociation constant for intrinsic biotin-binding affinity of mSA (fitted).

$$\theta_{bound} = A_1 \frac{-\left[L_t + [mSA] + \frac{K_c}{K_p}(P_t - L_t)\right] + \left[\left(L_t + [mSA] + \frac{K_c}{K_p}(P_t - L_t)\right)^2 + 4L_t[mSA]\frac{K_c}{K_p}\right]^{1/2}}{2P_o} + A_2 \quad (2)$$

P_t: Total concentration of Qsy7-labeled mSA (fixed); L_t: Total concentration of Oregon Green488-labeled SN-BAP-GpATM (fixed); A_1, A_2: Amplitude of binding phase and initial fluorescence level, respectively (fitted); [mSA]: Total concentration of unlabeled monovalent streptavidin (variable); K_c: Intrinsic biotin-binding affinity of unlabeled mSA (only fitted when K_p is known and fixed); K_p: Intrinsic biotin-binding affinity of Qsy7-labeled mSA (only fitted when K_c is known and fixed).

4.2 Preparation of Proteoliposomes

1. Add 15.2–60.8 mg of palmitoyloleoylphosphatidylcholine ($C_{16:0}C_{18:1c9}$PC, Avanti polar lipids) dissolved in chloroform (25 mg/ml) to glass tubes and dry under a stream of nitrogen gas. Remove residual solvent in a vacuum desiccator for 2 h.

2. Hydrate and solubilize the dried lipid films in 2 ml of 2–6 % β-octyl glucoside (Anatrace, Anagrade), 20 mM MOPS, 200 mM NaCl (pH 7.5). Add the DM solubilized SN-BAP-GpATM stock solution to the solubilized $C_{16:0}C_{18:1c9}$PC at a final concentration of 10 μM and incubate at room temperature for 1 h.

3. Reconstitute the SN-BAP-GpATM into $C_{16:0}C_{18:1c9}$PC vesicles by dialyzing against 250 sample volumes of 20 mM MOPS, 200 mM NaCl buffer solution (pH 7.5), with three equivalent buffer exchanges over the course of 48 h at 5 °C (Spectra/Por 25 kDa cutoff dialysis membrane, 7.5 mm diameter).

4. Extrude the resulting proteoliposomes through Nucleo track etch membrane with 200 nm pore size (Whatman) using a mini-extruder (Avanti polar lipids) 15 times. The typical average diameter of proteoliposomes is 110 ± 60 nm as measured by dynamic light scattering (DynaPro light-scattering systems, Protein Solutions). Store the liposomal solutions at 4 °C.

4.3 Determination of Lipid-to-Protein Molar Ratio

1. Solubilize the proteoliposomes prepared above with an equal volume of 200 mM C_8E_5 solution (20 mM MOPS, 200 mM NaCl, pH 7.5).

4.3.1 Determination of Total Protein and Lipid Concentration After Reconstitution

2. To determine SN-BAP-GpATM concentration, titrate the solubilized sample with wild-type mSA and incubate overnight.

3. Determine the total concentration of pyrene-labeled biotinylated SN-BAP-GpATM by fitting the stoichiometric binding data to Eq. 1 with a fixed parameter, $K_{d,biotin} = 10^{-14}$ M (Fig. 6a).

4. Measure the total lipid concentration in the proteoliposome samples by an organic-phosphate assay [22] (Fig. 6b).

4.3.2 Determination of the Orientation Distribution of SN-BAP-GpATM in Lipid Vesicles

The purpose of this procedure is to determine the effective protein concentration whose biotin tag is accessible to mSA from the outside of the vesicles. The orientation distribution of biotinylated SN-BAP-GpATM in $C_{16:0}C_{18:1c9}$PC bilayers is determined by an avidin-binding assay [11] (Fig. 6c).

1. Incubate the proteoliposome samples (~10 μM SN-BAP-GpATM) with a molar excess of avidin (Pierce) (20 μM in tetramer) overnight.

2. Add excess free biotin (1 mM) to saturate the rest of biotin-binding sites, thereby freezing the amount of avidin–SN-BAP-GpATM complex because the off-rate of biotin binding is extremely slow (see Fig. 3). Add SDS sample buffer and separate by SDS-PAGE without sample heating.

3. As a positive control for full binding, first solubilize the proteoliposomes in SDS sample buffer and then add excess avidin and free biotin sequentially as in **step 2** (Fig. 6c, lane 4). Avidin-accessible SN-BAP-GpATM is separated from the SN-BAP-GpATM band.

4. As a negative control, pre-incubate avidin with an excess free biotin prior to addition to the proteoliposomes. Add SDS sample buffer and separate by SDS-PAGE without sample heating (Fig. 6c, lane 2).

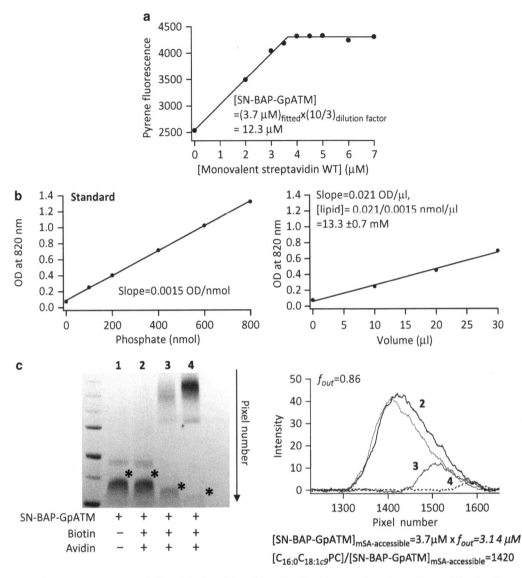

Fig. 6 Method and representative data for determining the lipid-to-protein molar ratio and the fraction of mSA-accessible SN-BAP-GpATM in lipid bilayers. Reproduced from Hong et al. [11]. (**a**) Measuring the total concentration of the biotinylated SN-BAP-GpATM by stoichiometric titration with wild-type mSA. (**b**) Organic-phosphate assay to measure the concentration of $C_{16:0}C_{18:1c9}$PC. (**c**) Determining the orientation distribution of the biotinylated SN-BAP-GpATM in $C_{16:0}C_{18:1c9}$PC bilayers by avidin-binding assay (*see* Subheading 3.4, **step 2**)

5. Process the digitized gel images using the LabView program (National Instruments) developed by Dr. Volker Kiessling in Professor Lukas Tamm lab at University of Virginia.

6. Digitized band intensities are background-subtracted and integrated using the Igor-Pro program (WaveMetrics) (Fig. 6c, right).

7. Determine the mol fraction of SN-BAP-GpATM with accessible biotin residues using the equation $f_{out} = [I(\text{Negative})$

control) − I(Experiment)]/[I(Negative control) − I(Positive control)]. Each term represents the integrated intensity of each SN-BAP-GpATM band, which is not bound to avidin.

8. Calculate the lipid-to-protein molar ratio (L/P) based on the equation, $L/P = C_{C16:0C18:1c9PC,total}/(C_{GpA,total}f_{out})$, in which $C_{C16:0C18:1c9PC,total}$, $C_{GpA,total}$, and f_{out} represent the total $C_{16:0}C_{18:1c9}PC$ concentration, total SN-BAP-GpATM concentration, and fraction of avidin-accessible SN-BAP-GpATM in a proteoliposome preparation, respectively. The mol fraction of avidin-accessible SN-BAP-GpATM is defined simply by $1/(L/P)$ since the protein contribution is negligible.

4.3.3 Measurement of mSA Binding

Binding of mSA with pyrene-labeled SN-BAP-GpATM is monitored by the increase in the fluorescence intensity at increasing concentrations of mSA (Fig. 7, left). The raw fluorescence changes

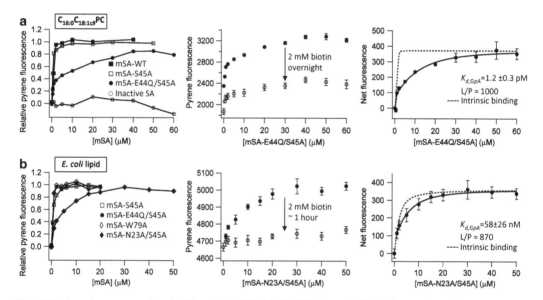

Fig. 7 Strategy to measure the stability of the SN-BAP-GpATM dimer in lipid bilayers. (a) Left: selection of a mutant mSA with an optimal intrinsic biotin affinity ($K_{d,biotin}$) to measure the high stability of the GpATM dimer in neutral $C_{16:0}C_{18:1c9}PC$ bilayers. The affinities of wild-type mSA ($K_{d,biotin} \sim 10^{-14}$ M) and mSA-S45A ($K_{d,biotin} = 2.1 \times 10^{-9}$ M) were so strong that the second binding phase could not be discriminated. mSA-E44Q/S45A ($K_{d,biotin} = 9.1 \times 10^{-9}$ M) yielded the desired biphasic curve with strong first binding followed by an attenuated second binding. Middle: Addition of excess free biotin induced the suppression of the fluorescence changes implying the reversibility of the GpATM dimerization. The raw fluorescence changes (filled circle) were background subtracted by the free biotin-added data (open circle). Right: The corrected curve was fitted to the Eq. 3, which yielded the apparent dissociation constant ($K_{d,GpA}$) of the GpATM dimer of $1.2 \pm 0.3 \times 10^{-12}$ M. Modified from Hong et al. [11]. (b) GpATM forms a weaker dimer in lipid bilayers composed of E. coli lipids. mSA-N23A/S45A with a weaker biotin affinity ($K_{d,biotin} = 8.7 \times 10^{-7}$ M) yielded the biphasic binding curve (right). The raw fluorescence changes were background subtracted by the data after the addition of free biotin (middle). Equation 4, which selectively fits the second binding phase, was used to obtain the $K_{d,GpA}$ of $4.8 \pm 2.0 \times 10^{-8}$ M. Modified and reprinted with permission from Hong and Bowie [26]. Copyright 2011 American Chemical Society

are subtracted by the data obtained after the addition of excess free biotin, which leads to a complete dissociation of the mSA-dissociated SN-BAP-GpATM complex (Fig. 7, middle). To obtain the thermodynamic stability of the GpATM dimer, an mSA mutant must be chosen from the mSA library that yields a characteristic biphasic curve (Figs. 1b, 2, and 7, left). Practically, the optimal mSA is determined based on the criteria that the weaker second mSA-binding phase can be fully observed at concentrations below 100 μM mSA. The corrected binding curve is fitted to the appropriate equations to extract the dissociation constant of the dimer ($K_{d,dimer}$) (Fig. 7, right; Eqs. 3 and 4).

1. For binding measurements in $C_{16:0}C_{18:1c9}PC$ proteoliposomal solution, mix 30 μl of proteoliposomes with various compositions of buffer/BSA (bovine serum albumin, Sigma)/mSA mixtures. Add BSA to match the total protein concentrations of the sample solutions as changes in protein concentration can affect osmolarity and viscosity and alter the observed fluorescence. Perform all measurements in 20 mM MOPS, 200 mM NaCl buffer solution (pH 7.5) at room temperature.

2. Adjust the total molar concentration of BSA and mSA to 100 μM in all sample solutions. Transfer a total volume of 100 μl of each sample to a microplate, sealed with a polyolefin tape, and incubated overnight at room temperature (see **Note 1**).

3. Carefully stir the sample solutions in the wells to resuspend settled proteoliposomes before measurements. Measure fluorescence in microtiter plate reader (Molecular Devices M5 model) using an excitation wavelength of 330 nm and an emission wavelength 390 nm. Measure and average the mSA-induced changes in the pyrene-fluorescence intensity at least five times.

4. At the end of the measurements, add excess free biotin to a final concentration of 2 mM and incubate more than 6 h at room temperature to dissociate bound mSA from the biotinylated SN-BAP-GpATM. The fluorescence data from the biotin-blocked samples serve as a background (see **Note 1**).

5. Fit the background-subtracted data to Eq. 3 to extract $K_{d,dimer}$ of the GpATM dimer.

$$\theta_{bound} = A_1 \frac{\left(P_o + [mSA] + K_{d,biotin}\right) - \left(\left(P_o + [mSA] + K_{d,biotin}\right)^2 - 4P_o[mSA]\right)^{1/2}}{2P_o}$$

$$+ A_2 \frac{-\left(1 + \dfrac{K_{d,biotin}}{[mSA]}\right) + \left(\left(1 + \dfrac{K_{d,biotin}}{[mSA]}\right)^2 + 8P_o\dfrac{K_{d,biotin}^2}{K_{d,dimer}}\dfrac{1}{[mSA]^2}\right)^{1/2}}{4P_o \dfrac{K_{d,biotin}^2}{K_{d,dimer}}\dfrac{1}{[mSA]^2}} + A_3 \quad (3)$$

P_o: Total SN-BAP-GpATM concentration (fixed); A_1,A_2,A_3: Amplitude of the first binding phase, amplitude of the second binding phase, and initial fluorescence level, respectively (fitted); [mSA]: Total mSA concentration (variable); $K_{d,biotin}$: Dissociation constant for intrinsic biotin-binding affinity of mSA (fixed); $K_{d,dimer}$: Dissociation constant for GpATM dimers (fitted).

$$\theta_{bound} = A_1 \frac{-\left(1 + \dfrac{K_{d,biotin}}{[mSA]}\right) + \left(\left(1 + \dfrac{K_{d,biotin}}{[mSA]}\right)^2 + 8P_o \dfrac{K_{d,biotin}^{\;2}}{K_{d,dimer}} \dfrac{1}{[mSA]^2}\right)^{1/2}}{4P_o \dfrac{K_{d,biotin}^{\;2}}{K_{d,dimer}} \dfrac{1}{[mSA]^2}} + A_2 \qquad (4)$$

P_o: Total SN-BAP-GpATM concentration (fixed); A_1,A_2: Amplitude of binding phase and initial fluorescence level, respectively (fitted); [mSA]: Total mSA concentration (variable); $K_{d,biotin}$: Dissociation constant for intrinsic biotin-binding affinity of mSA (fixed); $K_{d,dimer}$: Dissociation constant for GpATM dimers (fitted).

4.4 Conversion to Mol-Fraction Scale

The stability of TM helix dimers directly depends on the number of TM helices relative to the available hydrophobic volume (V_L) of lipid environments rather than on the number of the TM helices in the total solution volume ($V_L + V_{water}$) [23]. Consequently, the dimer dissociation constants should be expressed on a mol-fraction scale [23, 24].

$$K_{d,dimer}(X) = \frac{K_{d,dimer} \text{ in molar}}{[L] \text{ in molar}} \qquad (5)$$

$$\Delta G_X^o = RT \ln\left(K_{d,dimer} / [L]\right) = \Delta G^o - RT\ln[L] \qquad (6)$$

[L]: total molar concentration of lipid or detergent molecules; $K_{d,dimer}(X)$: dissociation constant in mol-fraction scale; $K_{d,dimer}$: apparent dissociation constant in molar scale; ΔG_X^o: standard free energy change of dimer dissociation in mol-fraction scale; ΔG^o: standard free energy change of dimer dissociation in molar scale.

To directly compare the free energy changes of dimer dissociation measured in different lipid environments, it is useful to plot the ΔG_X^o as a function of natural log of mol fraction of TM helices [11] (Fig. 8). If TM helices solubilized in the hydrophobic phase behave like ideal solute-solvent or if the dilution effect of the TM helix is purely entropic, the ΔG_X^o should be independent of the mol fraction of TM helix. This is the case for the dimerization of GpATM in C_8E_5 detergent micelles [23]. However, in decyl maltoside micelles and $C_{16:0}C_{18:1c9}PC$ lipid bilayers, ΔG_X^o shows a significant deviation from the ideal behavior, which indicates a complex

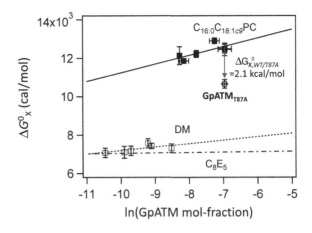

Fig. 8 Comparison of ΔG_X^o in detergent micelles (DM [11, 25] and C_8E_5 [23]) and $C_{16:0}C_{18:1c9}$PC bilayers as a function of mol fraction of SN-BAP-GpATM [11]. The mutational effect of the dimerization of GpATM can be evaluated by obtaining the difference in free energy changes ($\Delta\Delta G_{X,\,\text{wild-type/mutant}}^o$) between wild-type and GpATM mutant. The $\Delta\Delta G_{X,\,\text{wild-type/T87A}}^o$ was estimated from the difference between the experimentally determined $\Delta G_{X,\text{T87A}}^o$ at a given mol fraction and the $\Delta G_{X,\text{wild-type}}^o$ linearly interpolated to the same mol fraction. Modified from Hong et al. [11]

enthalpic and entropic contribution from the GpATM-detergent (or lipid) interactions in the monomer–dimer equilibrium or the involvement of higher order oligomers [23, 25]. Consequently, changes in dimerization free energy upon mutation or lipid composition changes can depend on the total protein concentration. To compare changes at a common concentration, the $\Delta G_X^o - \ln(\text{mol-fraction})$ plot can be used to obtain the difference free energy changes between wild-type and mutant ($\Delta\Delta G_{X,\text{wild-type/mutant}}^o = \Delta G_{X,\text{wild-type}}^o - \Delta G_{X,\text{mutant}}^o$) by comparing $\Delta G_{X,\text{mutant}}^o$ and the linearly interpolated values $\Delta G_{X,\text{wild-type}}^o$ at a given mol fraction of the TM helix [11] (*see* Fig. 8).

5   Conclusion and Outlook

The steric trap method described here has been successfully applied to study the stability of the GpATM dimer in lipid bilayers. Notably, the method could be used to measure dimer interaction at the sub-picomolar range in neutral fluid $C_{16:0}C_{18:1c9}$PC bilayers, which represents the strongest TM helix interaction ever measured. The method further revealed that the GpATM dimer is more tightly packed in lipid bilayers than in detergent micelles [11], and the lipid and protein composition of natural cell membranes dramatically destabilize the TM helix interaction [26]. The ultimate goal of this method is to tackle the folding and stability of larger α-helical membrane proteins in native bilayer environment. The tools established from the simple TM helix model system will provide a foundation for these more complex systems.

6   Note

1. We found that prolonged incubation for the weaker binding mutants led to aggregation, so it is important to try and minimize incubation times while still achieving equilibrium. The intrinsic rates of the dimer dissociation as well as the off-rates of the bound mSA mutants after the addition of excess free biotin are important factors to determine the incubation times in the thermodynamic binding and the background measurements, respectively. Practically, for the strong GpATM helix interactions, which are probed by strong biotin binding mSA mutants (mSA-S45A, mSA-E44Q/S45A, and mSA-W79M), it takes overnight for the system to reach equilibrium in mSA-binding measurements. To completely dissociate the SN-BAP-GpATM/mSA complex by addition of excess free biotin, it also takes at least 6 h of incubation. For the weaker interactions that can be probed by mSA mutants with weaker biotin affinities (mSA-W79Q, mSA-N23A/S45A, and mSA-S27R), the equilibrium for mSA binding is reached within 3 h, and the background fluorescence intensities are stabilized within 1 h after the free biotin addition (Fig. 7b).

Acknowledgements

We thank the Karen Fleming lab (Johns Hopkins University) and the Alice Ting lab (MIT) for providing plasmids. This work was supported by National Institutes of Health (NIH) Grants R01GM063919 and R01GM081783 (to J.U.B.) and start-up funds from Michigan State University (to H.H.). H.H. was supported by the Leukemia and Lymphoma Society Career Development Program (Fellow).

Appendix: Derivation of the Equation for the Second mSA Binding Coupled to the Dissociation of the TM Helix Dimer

Equation 4 was derived as follows according to the reaction scheme,

$$P_2 \rightleftarrows 2P \rightleftarrows 2P \cdot \text{mSA}$$

$$K_{d,\text{dimer}} = \frac{[P]^2}{[P_2]}, \quad [P_2] = \frac{[P]^2}{K_{d,\text{dimer}}} \quad (7)$$

$$K_{d,biotin} = \frac{[P][mSA]}{[P \cdot mSA]}, \quad [P \cdot mSA] = \frac{[P][mSA]}{K_{d,biotin}} \qquad (8)$$

$$\theta_{bound} = \frac{[P \cdot mSA]}{[P]_o} = \frac{[P \cdot mSA]}{2[P_2] + [P] + [P \cdot mSA]} = \frac{\dfrac{[P][mSA]}{K_{d,biotin}}}{2\dfrac{[P]^2}{K_{d,dimer}} + [P] + \dfrac{[P][mSA]}{K_{d,biotin}}}$$

$$(7) \wedge (8) = \frac{1}{1 + \dfrac{K_{d,biotin}}{[mSA]} + 2\dfrac{K_{d,biotin}}{K_{d,dimer}}\dfrac{[P]}{[mSA]}}$$

By using Eq. 8

$$[P] = \frac{K_{d,biotin}[P \cdot mSA]}{[mSA]} = \frac{K_{d,biotin}([P]_o \cdot \theta_{bound})}{[mSA]}$$

$$\theta_{bound} = \frac{1}{1 + \dfrac{K_{d,biotin}}{[mSA]} + 2\dfrac{K_{d,biotin}^2([P]_o \cdot \theta_{bound})}{K_{d,dimer}[mSA]^2}},$$

which gives

$$2\frac{K_{d,biotin}^2}{K_{d,dimer}}\frac{[P]_o}{[mSA]^2}\theta_{bound}^2 + \left(1 + \frac{K_{d,biotin}}{[mSA]}\right)\theta_{bound} - 1 = 0$$

Then,

$$\theta_{bound} = \frac{-\left(1 + \dfrac{K_{d,biotin}}{[mSA]}\right) + \left(\left(1 + \dfrac{K_{d,biotin}}{[mSA]}\right)^2 + 8P_o\dfrac{K_{d,biotin}^2}{K_{d,dimer}}\dfrac{1}{[mSA]^2}\right)^{1/2}}{4P_o\dfrac{K_{d,biotin}^2}{K_{d,dimer}}\dfrac{1}{[mSA]^2}}$$

$[P]$: SN-BAP-GpATM monomer concentration; $[P_2]$: SN-BAP-GpATM dimer concentration; $[P]_o$: Total SN-BAP-GpATM concentration; $[mSA]$: Total mSA concentration; $K_{d,biotin}$: Dissociation constant for intrinsic biotin-binding affinity of mSA; $K_{d,dimer}$: Dissociation constant for GpATM dimers.

References

1. Moore DT, Berger BW, DeGrado WF (2008) Protein-protein interactions in the membrane: sequence, structural, and biological motifs. Structure 16:991–1001

2. MacKenzie KR, Fleming KG (2008) Association energetics of membrane spanning alpha-helices. Curr Opin Struct Biol 18: 412–419

3. Senes A (2011) Computational design of membrane proteins. Curr Opin Struct Biol 21: 460–466

4. Fisher LE, Engelman DM, Sturgis JN (1999) Detergents modulate dimerization, but not helicity, of the glycophorin A transmembrane domain. J Mol Biol 293:639–651

5. You M, Li E, Wimley WC, Hristova K (2005) Forster resonance energy transfer in liposomes: measurements of transmembrane helix dimerization in the native bilayer environment. Anal Biochem 340:154–164

6. Cristian L, Lear JD, DeGrado WF (2003) Use of thiol-disulfide equilibria to measure the energetics of assembly of transmembrane helices in phospholipid bilayers. Proc Natl Acad Sci USA 100:14772–14777

7. Fleming KG, Ackerman AL, Engelman DM (1997) The effect of point mutations on the free energy of transmembrane alpha-helix dimerization. J Mol Biol 272:266–275

8. Hong H, Joh NH, Bowie JU, Tamm LK (2009) Methods for measuring the thermodynamic stability of membrane proteins. Methods Enzymol 455:213–236

9. Chen L, Novicky L, Merzlyakov M, Hristov T, Hristova K (2010) Measuring the energetics of membrane protein dimerization in mammalian membranes. J Am Chem Soc 132:3628–3635

10. Blois TM, Hong H, Kim TH, Bowie JU (2009) Protein unfolding with a steric trap. J Am Chem Soc 131:13914–13915

11. Hong H, Blois TM, Cao Z, Bowie JU (2010) Method to measure strong protein-protein interactions in lipid bilayers using a steric trap. Proc Natl Acad Sci USA 107:19802–19807

12. Howarth M, Chinnapen DJ, Gerrow K, Dorrestein PC, Grandy MR, Kelleher NL, El-Husseini A, Ting AY (2006) A monovalent streptavidin with a single femtomolar biotin binding site. Nat Methods 3:267–273

13. Laitinen OH, Hytonen VP, Nordlund HR, Kulomaa MS (2006) Genetically engineered avidins and streptavidins. Cell Mol Life Sci 63:2992–3017

14. Lemmon MA, Flanagan JM, Hunt JF, Adair BD, Bormann BJ, Dempsey CE, Engelman DM (1992) Glycophorin A dimerization is driven by specific interactions between transmembrane alpha-helices. J Biol Chem 267: 7683–7689

15. Fleming KG, Engelman DM (2001) Specificity in transmembrane helix-helix interactions can define a hierarchy of stability for sequence variants. Proc Natl Acad Sci USA 98: 14340–14344

16. Beckett D, Kovaleva E, Schatz PJ (1999) A minimal peptide substrate in biotin holoenzyme synthetase-catalyzed biotinylation. Protein Sci 8:921–929

17. Orzaez M, Perez-Paya E, Mingarro I (2000) Influence of the C-terminus of the glycophorin A transmembrane fragment on the dimerization process. Protein Sci 9:1246–1253

18. Wu SC, Wong SL (2004) Development of an enzymatic method for site-specific incorporation of desthiobiotin to recombinant proteins in vitro. Anal Biochem 331:340–348

19. Chilkoti A, Tan PH, Stayton PS (1995) Site-directed mutagenesis studies of the high-affinity streptavidin-biotin complex: contributions of tryptophan residues 79, 108, and 120. Proc Natl Acad Sci USA 92:1754–1758

20. Doura AK, Kobus FJ, Dubrovsky L, Hibbard E, Fleming KG (2004) Sequence context modulates the stability of a GxxxG-mediated transmembrane helix-helix dimer. J Mol Biol 341:991–998

21. Hyre DE, Le Trong I, Freitag S, Stenkamp RE, Stayton PS (2000) Ser45 plays an important role in managing both the equilibrium and transition state energetics of the streptavidin-biotin system. Protein Sci 9:878–885

22. Ames BN, Dubin DT (1960) The role of polyamines in the neutralization of bacteriophage deoxyribonucleic acid. J Biol Chem 235: 769–775

23. Fleming KG (2002) Standardizing the free energy change of transmembrane helix-helix interactions. J Mol Biol 323:563–571

24. White SH, Wimley WC (1999) Membrane protein folding and stability: physical principles. Annu Rev Biophys Biomol Struct 28: 319–365

25. Fisher LE, Engelman DM, Sturgis JN (2003) Effect of detergents on the association of the glycophorin a transmembrane helix. Biophys J 85:3097–3105

26. Hong H, Bowie JU (2011) Dramatic destabilization of transmembrane helix interactions by features of natural membrane environments. J Am Chem Soc 133:11389–11398

Chapter 4

Genetic Systems for Monitoring Interactions of Transmembrane Domains in Bacterial Membranes

Lydia Tome, Dominik Steindorf, and Dirk Schneider

Abstract

In recent years several systems have been developed to study interactions of TM domains within the inner membrane of the Gram-negative bacterium *Escherichia coli*. Mostly, a transmembrane domain of interest is fused to a soluble DNA-binding domain, which dimerizes in *E. coli* cytoplasm after interactions of the transmembrane domains. The dimeric DNA-binding domain subsequently binds to a promoter/operator region and thereby activates or represses a reporter gene. In 1996 the first bacterial system has been introduced to measure interactions of TM helices within a bacterial membrane, which is based on fusion of a transmembrane helix of interest to the DNA-binding domain of the *Vibrio cholerae* ToxR protein. Interaction of a transmembrane helix of interest within the membrane environment results in dimerization of the DNA-binding domain in the bacterial cytoplasm, and the dimeric DNA-binding domain then binds to the DNA and activates a reporter gene. Subsequently, systems with improved features, such as the TOXCAT- or POSSYCCAT system, which allow screening of TM domain libraries, or the GALLEX system, which allows measuring heterotypic interactions of TM helices, have been developed and successfully applied. Here we briefly introduce the currently most applied systems and discuss their advantages together with their limitations.

Key words BACTH, Dimerization, GALLEX, Genetic system, Helix–helix interaction, Transmembrane helix, TOXCAT

1 Introduction

Due to their hydrophobic nature, investigation of membrane-integrated proteins is a difficult task [1]. Typically, transmembrane (TM) proteins are analyzed after extraction from membranes, and the choice of a proper detergent for the extraction and purification is crucial for subsequent analysis, and sometimes usage of the wrong detergent can even result in inactivation and/or denaturation of the protein. As the detergent environment might severely influence the structure of TM proteins, it is desirable to analyze interactions of TM proteins in vivo, i.e., within a biological

Lydia Tome and Dominik Steindorf have contributed equally to this work.

Giovanna Ghirlanda and Alessandro Senes (eds.), *Membrane Proteins: Folding, Association, and Design*, Methods in Molecular Biology, vol. 1063, DOI 10.1007/978-1-62703-583-5_4, © Springer Science+Business Media, LLC 2013

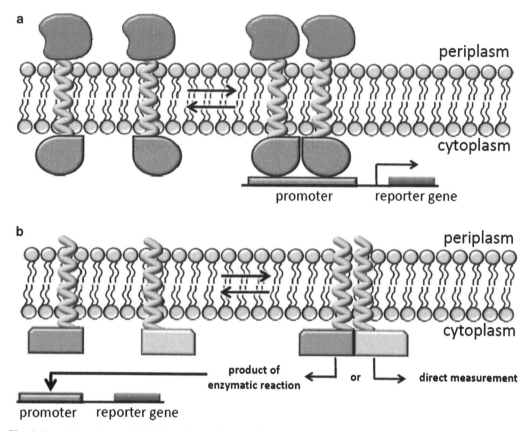

Fig. 1 How interactions of TM domains can be monitored in the *E. coli* inner membrane. (**a**) Activation of a reporter gene can be mediated by a DBD, which is genetically fused to the TM domain. A chimeric protein containing a TM domain of interest and a cytosolic DBD is expressed into the *E. coli* inner membrane. Upon interaction of the TM domains, the DBD binds to a promoter/operator region, resulting in a reporter gene activation or repression. (**b**) Measuring interaction of TM domains by enzyme fragment complementation. An enzyme is split into two domains, which can be separately expressed, and both exhibit no enzymatic activity. The enzymatic activity is restored when the two fragments are genetically fused to interacting TM domains and the TM domain interaction mediates enzyme complementation. The reconstituted enzymatic activity might be measured directly. Alternatively, products of the enzymatic reaction can activate a DNA-binding molecule that subsequently activates a reporter gene

membrane. In contrast to eukaryotic cells, analyses of TM protein interactions in a bacterial membrane have several advantages, as it is, for example, easily accessible, less time consuming, and highly reproducible. Because of this, in recent years several systems have been developed to study interactions of TM domains within the inner membrane of the Gram-negative bacterium *Escherichia coli*. In principle, interactions of TM helices can be monitored by two approaches (Fig. 1). One approach is based on the fusion of the TM helix of interest to a soluble DNA-binding domain (DBD), which dimerizes in *E. coli* cytoplasm after TM helix–helix interaction. The DBD dimer then binds to a promoter/operator and thereby activates or represses a reporter gene. In addition, an enzyme might be split into two halves, which are each fused to a TM helix. The individual halves are inactive and enzymatic activity is regained only when the two halves get into close contact caused by helix–helix association.

In 1996 Langosch and coworkers developed the first bacterial system to measure interactions of TM helices within a bacterial membrane, which is based on fusion of a TM helix of interest to the ToxR DBD [2]. Interaction of a TM helix of interest within the membrane environment results in dimerization of the ToxR DBD in the bacterial cytoplasm. The dimeric ToxR DBD then binds to the DNA and thereby activates a reporter gene. Subsequently, systems with improved features, such as the TOXCAT- [3] or POSSYCCAT system [4], which allow screening of TM domain libraries, or the GALLEX system [5], which allows measuring heterotypic interactions of TM helices, have been developed and successfully applied. In the following we briefly introduce the currently most applied systems and discuss their advantages together with their limitations.

1.1 Systems Based on the ToxR Transcription Activator: The ToxR-, TOXCAT-, and POSSYCCAT System

The ToxR protein of the bacterium *Vibrio cholerae* is a integral membrane protein, which contains an N-terminal DBD fused to a single TM helix and a C-terminal domain with a periplasmic localization. In vivo, ToxR is involved in transcription activation of cholera toxin genes. Dimerization of the periplasmatic ToxR domains triggers binding of the cytoplasmatic domains to DNA. ToxR is active and binds to DNA only as a homodimer and is able to also activate transcription starting at the *ctx* promoter in other bacteria, such as *E. coli* [6–8]. As the ToxR periplasmatic and TM domains are not intimately involved in DNA-binding, these can be replaced without losing the ability of the dimeric ToxR DBD to activate transcription at the *ctx* promoter [9]. Based on this observation, chimeric TM proteins have been designed by fusing the N-terminal ToxR DBD to a TM helix of interest (Fig. 2a [3]). Homooligomerization of a TM helix triggers homooligomerization of the ToxR DBD in the *E. coli* cytoplasm. The homooligomeric ToxR DBD can activate a *ctx* promoter fused to the *lacZ* reporter gene, which encodes the β-galactosidase. As homotypic interactions of the TM helices control reporter gene activity, the reporter gene activity, i.e., the amount of expressed reporter, directly reflects the strength of a TM helix–helix interaction (Table 1). Thus, a higher reporter gene expression and activity of the encoded reporter reflects a higher interaction propensity. Besides the ToxR DBD, the monomeric maltose-binding protein (MalE) of *E. coli* is fused to the C-terminus of the chimeric protein. This domain has been shown to facilitate membrane integration [9]. Noteworthy, fusing the MalE domain to the C-terminus of the chimeric protein has additional advantages, as it allows easy detection of the protein and determination of protein topology (*see* **Note 1**). Originally, expression of the designed chimeric proteins was under the control of the *toxR* promoter, resulting in

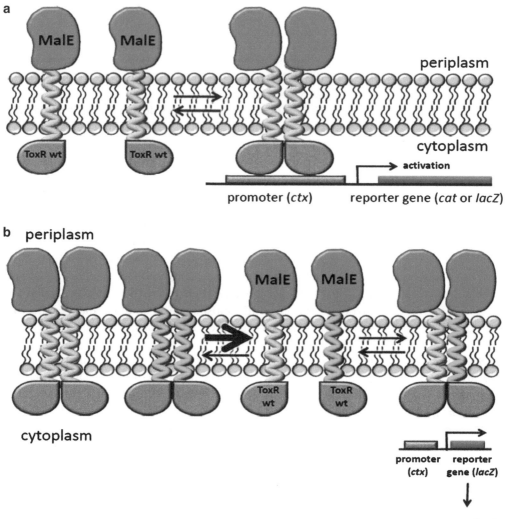

Fig. 2 ToxR-based systems. (**a**) Fusion proteins are integrated into the *E. coli* inner membrane with the MalE protein localized in the *E. coli* periplasm and the ToxR domain in the cytoplasm. Activation of a reporter gene (either *lacZ* or *cat*) is driven by binding of the dimerized ToxR DBD to the *ctx* promoter. (**b**) Positive ToxR assay: Two different TM domains are genetically fused to the soluble wt ToxR DBD and are separately expressed from different plasmids. Homotypic interaction of one of the two TM helices results in weak activation of the *lacZ* reporter gene. In case of a strong heterotypic interaction of the two expressed TM domains, the equilibrium is significantly shifted towards the heterodimer, which leads to strong reporter gene activation. This variant of the ToxR assay is only suitable to monitor heterotypic interactions, which are stronger than the corresponding homotypic interactions. (**c**) A dominant-negative ToxR assay: Two chimeric proteins with different TM domains are expressed from two different plasmids. One of the TM domains is genetically fused to the inactive ToxR S87H DBD whereas the other chimeric protein contains the wt ToxR DBD. Dimerization of the inactive ToxR S87H variant, mediated by homotypic interaction of fused TM domains, does not activate the *lacZ* gene, whereas dimerization of the wt ToxR DBD results in reporter activation. Heterotypic interaction of the expressed TM domains results in formation of wt/S87H ToxR heterodimers, resulting in weak or no activation of the reporter gene. (**d**) Inhibition of ToxR DBD dimer formation by externally added TM peptides. Activation of the *lacZ* reporter gene is mediated by the ToxR DBD dimer. The ToxR DBD is genetically fused to an interacting TM domain. Dimerization of the ToxR DBD, and thus activation of the reporter gene expression, can be inhibited or diminished by externally adding a synthetic TM peptide, which interacts with the ToxR DBD-containing chimeric protein

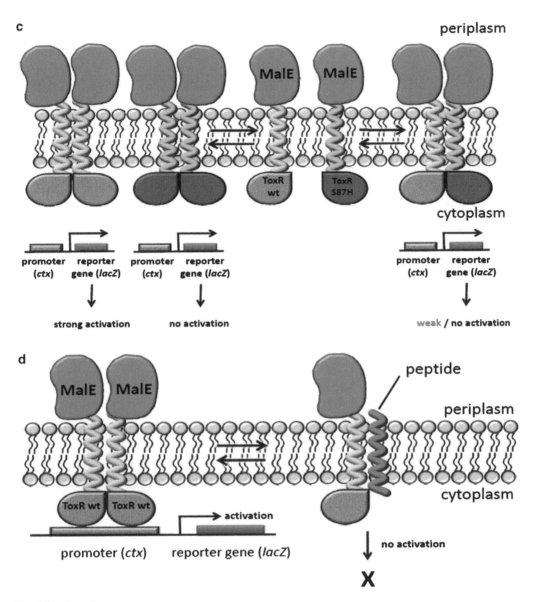

Fig. 2 (continued)

constitutive protein expression. As a proof of concept, dimerization of the well-characterized glycophorin A (GpA) TM helix has initially been analyzed in the *E. coli* inner membrane with the ToxR system by measuring β-galactosidase activities of the GpA wt and of mutant TM helices (*see* **Notes 2** and **3**) [2, 9, 10]. As expression of chimeric proteins from the *toxR* promoter cannot be controlled, and thus, as expression controlled by this promoter might alter under different conditions, advanced plasmids (pToxRIV and pToxRV) have been developed, which allow protein expression to be controlled by the inducible *ara* promoter (*see* **Note 4**) [11–15].

In 1999 Engelman and coworkers have introduced the TOXCAT system, another ToxR-based system, which allows

Table 1

Genetic systems to measure TM helix–helix interactions and their relevant features

System name	ToxR, classical	ToxR, dominant-negative	ToxR, positive	TOXCAT	POSSYCCAT	GALLEX	BACTH
Type of interaction measured	Homodimerization	Homodimerization–heterodimerization comparison	Homodimerization–heterodimerization comparison	Homodimeri-zation	Same as ToxR systems except TOXCAT	Homo- or heterodimerization	Homo- or heterodimerization
Orientation of transmembrane helices	Parallel	Parallel	Parallel	Parallel	Parallel	Parallel	Parallel or antiparallel
E. coli reporter strain(s)	FHK12	FHK12	FHK12	NT326	EL81, EL83 (ToxR assay) Chr3, EL47, EL61, EL141 (selection)	SU101 (homodimerization) SU202 (heterodimerization)	DHM1 or BTH101
Reporter gene(s)	*lacZ*	*lacZ*	*lacZ*	*cat*	*lacZ, cat, tet^R* (promoters of reporter genes: *ctx, ompT, ompU*)	*lacZ*	*lacZ*
Reporter gene localization	Genome	Genome	Genome	Plasmid	Genome	Genome	Genome
Reporter gene activation or repression	Direct *activation* by DBD	Direct *activation* by DBD	Direct *activation* by DBD	Direct *activation* by DBD	Direct *activation* by DBD	Direct *repression* by DBD	Indirect *activation* by cAMP/CAP complex
DNA-binding domain	ToxR from *V. cholerae*	ToxR from *V. cholerae*	ToxR from *V. cholerae*	ToxR from *V. cholerae*	ToxR from *V. cholerae*	LexA from *E. coli*; LexA 408	None, reporter gene activation through cAMP/CAP complex
Promoter controlling chimeric protein expression	*toxR* (constitutive)/ *ara* (arabinose inducible)	*toxR/ara*	*toxR/ara*	*toxR*	*toxR/ara*	*tac*	*lacUV5* (cAMP/CAP independent)
Plasmids (*see* Tables 4 and 6)	pToxRIV, pToxRV	pToxRVII/ pToxRVIII, pToxRVI	pToxRVII, pToxRIX	pccKan	All ToxR-based system plasmids except pccKan	pALM148, pBLM100	pKT25/pKNT25, pUT18C/pUT18

investigating homooligomerization propensities of TM helices [16–20]. The expressed chimeric protein is identical as described above for the ToxR system. However, in contrast to the original ToxR system, in the TOXCAT system the *cat* reporter gene, encoding the chloramphenicol-acetyl-transferase (CAT), is fused to the *ctx* promoter. Thus, interaction of TM domains activates the *cat* reporter gene, and expression of CAT results in acetylation of chloramphenicol, which confers resistance to this antibiotic [3]. Furthermore, as increased interaction of TM helices, resulting in increased *cat* expression, refers resistance to chloramphenicol, increased *cat* expression enables *E. coli* cells to survive on increasing chloramphenicol concentrations. In library screens, randomized TM sequences were genetically fused to the ToxR DBD and the MalE domain, and sequences with a high homodimerization propensity have been selected on agar plates containing increasing chloramphenicol concentrations. Intriguingly, in a screen for new TM helix interaction sequences, sequences containing the (well-characterized) GxxxG-motif have been found to be highly overrepresented as well as serine- and threonine-containing motifs [21].

Also, in the POSSYCCAT system (*po*sitive *s*election *sy*stem based on *c*hromosomally integrated *CAT*) the *ctx* promoter controls expression of the *cat* reporter gene. However, in contrast to the TOXCAT system, the reporter gene is integrated in the genome of the *E. coli* strain Chr3 (*see* **Note 5**). Furthermore, in contrast to the TOXCAT system, expression of chimeric proteins from the pToxRIV plasmid is controllable by the arabinose-inducible *ara* promoter. With the POSSYCCAT system, in library screens, motifs mediating interactions of membrane-spanning leucine zippers have been analyzed [4]. Furthermore, the POSSYCCAT system has been extended by the development of new *E. coli* reporter strains, harboring different reporter genes (*lacZ*, *cat*, or *tet*) downstream of the *ompU* or *ompT* promoters, which are also targeted by the ToxR DBD [22]. The *E. coli* strain EL141 has been developed containing chromosomally integrated *ompT::tet* and *ompU::cat* promoter fusions. Binding of the ToxR DBD dimer to the *ompU* promoter leads to expression of the *cat* reporter gene, conferring resistance to chloramphenicol. Expression of the *tet* reporter gene downstream is repressed upon binding of the ToxR DBD dimer to the *ompT* promoter, resulting in reduced resistance to tetracycline. Using both available reports, TM helices have been identified, which display only moderate interaction propensities [22]. TM helices with weak interaction propensities were removed by positive selection by using the *ompU* reporter, as weak interactions did not trigger sufficient *cat* expression. By the concurrent use of the ompT reporter and addition of tetracycline, helices with high interaction propensities were removed due to negative selection, since strong interaction prevents tetracycline resistance.

Finally, the selected clones harbored TM helices with medium interaction propensities [4, 22].

The ToxR-based systems have also been applied to analyze heteroassociation of two different TM helices within the *E. coli* inner membrane, when homodimer formation of the individual TM helices is significantly weaker than heterodimer formation (Fig. 2b) [23]. To this end, two TM helices of interest, fused to the wt ToxR DBD, have been expressed from a high-copy-number plasmid and a low-copy-number plasmid. In order to assess the propensity to form homodimers, the helices have been individually expressed from each plasmid in the *E. coli* reporter strain and homooligomerization propensities have been determined. For the analysis of heterodimerization the two TM helices have to be coexpressed, with one chimeric protein expressed from the low-copy and the other one from the high-copy-number plasmid. Heterodimer formation is then quantified by comparing the mean β-galactosidase activities measured upon coexpressing both proteins to the mean β-galactosidase activity of cultures expressing solely the individual TM helices [23]. To be able to distinguish between the proteins derived from the high- and the low-copy-number plasmid, the his6-tag and the myc-tag coding regions have been removed in the low-copy-number plasmid, resulting in a lowered molecular mass of the expressed fusion proteins on SDS gels. Using this approach, Langosch and coworkers, e.g., analyzed the important role of oppositely charged residues for TM helix homo- and heterooligomerization [23].

In addition, the system has recently been modified to also analyze heterotypic interactions in a ToxR-based dominant-negative system (Fig. 2c). A homotypically interacting TM helix fused to the wt ToxR DBD is coexpressed together with a competitor protein fused to the inactive ToxR S87G mutant. While wt ToxR dimers strongly activate transcription of the *lacZ* reporter gene, heteromers of the wt ToxR and the mutant S87G ToxR DBD are essentially inactive [24]. Therefore, heteromer formation results in reduced β-galactosidase activity when compared to wt ToxR homooligomers, and the reduced reporter gene activity directly correlates with the tendency of the TM helices to heterooligomerize (*see* **Note 6**). Noteworthy, this dominant-negative system allows characterization of TM domain interactions only when homooligomerization competes with heterooligomerization.

A similar approach to analyze heterodimerization of TM helices with the ToxR system has been introduced by Shai and coworkers (Fig. 2d) [25]. Here, heteroassociation of a TM helix, genetically fused to MalE and ToxR DBD, and a chemically synthesized peptide has been analyzed with the ToxR system. When an expressed chimeric protein dimerizes and activates the *ctx* promoter, an externally added peptide, which interacts with the TM region of the expressed chimeric protein, interferes with TM helix homoassociation, and thus the *ctx* promoter is less activated and the β-galactosidase activity is lowered.

1.2 GALLEX

LexA is a transcription repressor, involved in the SOS response in bacteria by negatively regulating the activity of about 20 SOS genes. In vivo LexA consists of an N-terminal DBD and a C-terminal homodimerization domain [26, 27]. When the DNA of a cell is damaged or DNA replication is inhibited, LexA is inactivated by proteolytical cleavage between the two domains, leading to monomerization of the DBD and subsequent activation of SOS gene transcription [28]. As the C-terminal dimerization domain can be replaced by different interacting proteins or protein domains, the homodimerization propensities of proteins of interest can be analyzed in vivo [5, 29–32] Furthermore, a triple-mutated LexA variant (LexA408) has been identified, which recognizes an altered operator sequence (*op408*) compared to the wt protein. Thus, only a heterodimer composed of one wt LexA and the mutated LexA408 DBD can bind to a designed *op+/op408* promoter/operator, which allows monitoring heterotypic interactions of proteins fused to the respective DBD (Fig. 3). Thus, with the GALLEX system homo- as well as heterotypic interactions of TM α-helices can be monitored [5]. Similar to the ToxR-based systems, the TM helix of interest is fused at its N-terminus to the LexA DBD and C-terminally to the *E. coli* MalE protein (*see* **Note 1**). The hydrophobic TM helix acts as a membrane insertion signal, placing the LexA DBD in the cytoplasm and the MalE domain in the periplasm of *E. coli*. The stronger TM helices interact, the weaker expression of the β-galactosidase reporter gene (*lacZ*) is activated. Noteworthy, this is opposite to the above described ToxR-based systems, where activation and not repression of a reporter gene is measured. Thus, expression of non-interacting proteins in the GALLEX system results in high β-galactosidase activity due to the lack of reporter gene repression. For comparison, representative measurements of the reporter gene activities determined after analyzing the interaction propensities of wt and G83I-mutated GpA TM helices, measured with the ToxR-, TOXCAT-, and the GALLEX system, respectively, are shown in Fig. 6.

For measuring homooligomerization, a TM helix of interest, fused to a wild-type LexA DBD, has to be expressed in the *E. coli* SU101 strain. In this reporter strain, the reporter gene (*lacZ*) is under control of a wt LexA recognition sequence (*op+/op+*) [30]. If the TM α-helices interact, the LexA dimer will bind to the promoter/operator and repress *lacZ* transcription (Fig. 3a). For measuring interactions of two different α-helices, one α-helix has to be genetically fused to the wt LexA and the other to the LexA408 DBD (Fig. 3b). In the *E. coli* reporter strain SU202, a chromosomal copy of the *lacZ* reporter gene is under the control of an *op408/op+* hybrid promoter/operator, which is recognized only by the LexA408/LexAwt dimer [30] (*see* **Note 7**). The GALLEX system has initially been developed and used to monitor homo- and heterotypic interactions of single-span TM helices, such as integrins,

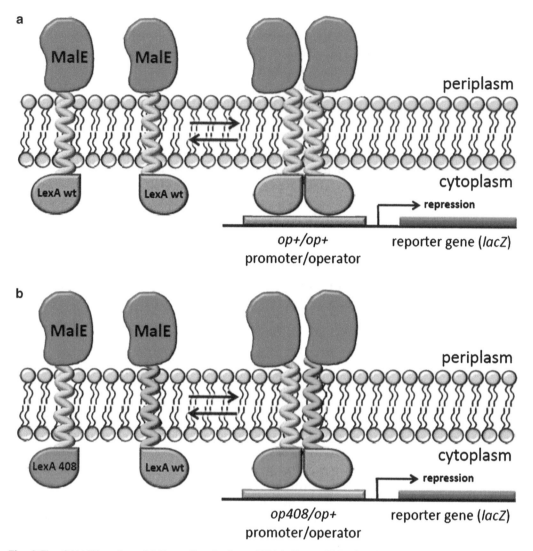

Fig. 3 The GALLEX system. (**a**) Homodimerization of TM helices within the *E. coli* inner membrane leads to formation of a wt LexA DBD dimer. DNA binding of this dimer to the *op+/op+* promoter/operator region results in repression of the *lacZ* reporter gene. (**b**) Measurement of heterodimeric interactions is possible by using an *E. coli* reporter strain SU202 carrying the *op408/op+* promoter/operator region. One of the expressed TM helices is genetically fused to the wt and a second TM helix to the Lex408 DBD. Only the heterodimeric LexA/Lex408 DBD dimer binds to the engineered *op408/op+* promoter/operator region, resulting in the repression of the *lacZ* reporter gene

ErbB receptor tyrosine kinases, or the TM helices of the major histocompatibility complex α and β subunits [31, 33, 34]. In addition, the GALLEX system has been applied to investigate a potential impact of cofactor binding on TM interaction, as measured with the TM b-type cytochrome b_{559}' [35]. Recently, also homotypic interactions of the multi-spanning *E. coli* glycerol facilitator GlpF and the diacylglycerol kinase have been monitored successfully with the GALLEX system [36]. Thus, the system allows not only monitoring interactions of isolated TM helices but also the observation of oligomerization of multi-span TM proteins. As expression of the fusion protein is controlled by the IPTG-inducible Ptac promoter in the GALLEX system, the level of the expressed chimeric protein can be gradually increased. Thermodynamically, the strength of a TM helix–helix interaction depends on the actual concentration of a TM helix within the membrane [37, 38]. In an experiment using the GpA TM helix as a model, a sigmoidal curve has been obtained when interaction propensities were determined at increasing IPTG and thus at increasing protein concentrations. By analyzing interaction propensities of TM helices in the presence of increasing IPTG concentrations, apparent K_D and ΔG values can be determined, and thus the GALLEX system allows estimating the energetics associated with mutations of single amino acids in TM domains (*see* **Note 8**). Nevertheless, while the GALLEX system allows monitoring the interaction propensity of two different helices/proteins, the analyzed TM domains still have to be oriented parallel to each other within the membrane (*see* **Note 9**).

1.3 BACTH

While not discussed in great detail here, the BACTH system ("*b*acterial *a*denylate *c*yclase *two* *h*ybrid") has emerged as another interesting possibility for studying interactions of TM helices/proteins. The BACTH system has initially been developed to measure interactions of proteins in the cytosol of *E. coli* cells [39–44]. The system is based on fragment complementation and restoration of an adenylate cyclase activity [45], and enzymatic conversion of adenosine triphosphate to cyclic adenosine monophosphate (cAMP) is measured (Fig. 4). In the BACTH system, the adenylate cyclase of *Bordetella pertussis* is split in the two complementary fragments T18 and T25, which can be separately expressed in *E. coli*. The enzymes' catalytic activity, and thus the ability to produce cAMP, is restored when the fragments are fused to interacting proteins. Produced cAMP binds to the catabolite activator protein (CAP), and binding of the cAMP/CAP complex to a promoter/operator region controls expression of reporter genes, organized, e.g., in the *mal* or *lac* operons in *E. coli* (*see* **Note 10**) [46–48]. Thus, in the BACTH system, functional interaction of the adenylate cyclase fragments, e.g., mediated by interacting TM domains, results in cAMP production (*see* **Note 11**). Although the BACTH system has initially been designed to investigate interactions of soluble proteins, it has been shown already that it is also capable to measure interactions of full-length TM proteins [49, 50]. Importantly, in contrast to the above described systems, it is possible to genetically fuse a TM helix of interest to both termini of the two respective adenylate cyclase fragments. Two plasmids for expressing T25 and T18 fusions have been engineered, containing a multiple cloning site upstream or downstream of the fragment's sequence, respectively (*see* **Note 12** [39, 49]). Thus, in principle formation of parallel TM domain homo- and herooligomers can

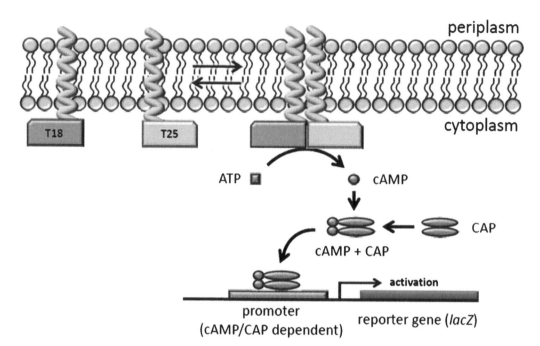

Fig. 4 The BACTH system. The BACT system is based on expression of two complementary and enzymatically inactive fragments of the *Bordetella pertussis* adenylate cyclase. A TM protein is genetically fused to the T18 and another TM protein to the T25 domain. When dimerization of the TM domains drives association of the isolated enzyme fragments, the enzyme's catalytic activity is restored and cAMP is synthesized. Formation of a cAMP/CAP complex activates expression of reporter genes by binding to a cAMP/CAP-dependent promoter region

be monitored with the BACTH system, as well as formation of antiparallel oligomers (*see* **Notes 1** and **13**).

1.4 Make the Right Choice: Features, Advantages, and Disadvantages of the Systems

For choosing the proper system to analyze interactions of TM domains, the individual characteristics of each system have to be considered (Table 1). ToxR-based systems offer the possibility to analyze homotypic interactions of TM domains, which form parallel oligomers within the membrane. Expression of the fusion proteins is either under control of the constitutive *toxR* promoter or the inducible *lac* promoter. The use of the *toxR* promoter does not allow adjusting the protein expression level, resulting in a low protein expression level that may vary under different growth conditions. In order to find the optimal assay conditions, it might be advantageous to test different expression levels of a fusion protein, and then it is recommended to use the vectors with the inducible *ara* promoter. In the early developed "classical" ToxR assay, dimerization of the ToxR DBD leads to activation of the *lacZ* reporter gene. While the activity of the β-galactosidase is rather easy to monitor and biochemical assays are established, in the TOXCAT and POSSYCCAT systems, *cat* is used as a reporter gene, conferring resistance to chloramphenicol. Depending on the strength of interaction, transformed *E. coli* clones are able to survive in media containing increasing chloramphenicol concentrations, which is suitable for library screening. Screening experiments are not possible when *lacZ* is used as a reporter gene. In addition, the POSSYCCAT system allows screening for TM helices with medium interaction propensities, because of the combined use of the *ompU* and *ompT* promoters. The modified dominant-negative system allows comparing the propensity of a TM helix to form homooligomers or heterooligomers with a competitive helix, respectively. This is possible due to the fusion of the competing helix to the mutated S87G ToxR DBD, leading to formation of an active homodimer as well as an inactive heterodimer. The ToxR system is also suitable for measuring heterodimerization of two TM helices, when the homooligomerization propensity is weaker than the heterodimerization propensity. In the vectors used in the GALLEX system that is suitable for measuring homo- as well as heterotypic interactions of parallel-oriented TM domains, protein expression is under control of the *tac* promoter. This allows estimation of energetics associated with TM domain interactions. Although addition of the inducer IPTG has a direct influence on the protein expression level, there always is a constitutive low background expression of the encoded chimeric protein, as the *lac* promoter is leaky [38].

Besides the mentioned advantages and the possibilities arising by having the different genetic systems, each system may have its limitation and several problems might arise and should be considered when the systems are used to analyze interactions of TM domains:

1. Fusion of TM helices to the soluble MalE domain facilitates membrane insertion, ensures the proper topology of the chimeric protein within the membrane, and allows assessing these properties with the maltose complementation assay (Subheading 3.8). However, the large MalE domain and the soluble cytoplasmic domains might affect proper interaction of TM helices due to steric clashes (*see* **Note 1**) [51].

2. Fusion of a TM domain to a DBD requires that the DBD is located in the bacterial cytoplasm. This can lead to a TM domain orientation different from its natural orientation, switching the N-terminal end from the cytoplasm to periplasm or vice versa. An altered TM topology might result in different interaction propensities.

3. Both, the ToxR-based systems and the GALLEX system have been shown to be sensitive to the length of a TM α-helix and to the relative orientation of the fused TM α-helix to the fusion domains (*see* **Note 9**) [2, 5].

4. Another limitation of the GALLEX and BACTH systems is the use of two different plasmids that differ in their respective copy number per cell, resulting in different protein expression levels

(*see* **Notes** **7** and **12**). When individual TM helices form homo- as well as heterooligomers, multiple monomer–oligomer equilibria are involved, and these equilibria are affected by the effective concentration of the expressed fusion proteins.

5. In contrast to the GALLEX system and the ToxR-based systems, in the BACTH system the reporter gene activity is not directly correlated to the strength of TM helix–helix interactions. In the GALLEX- and the ToxR-based systems, TM helix–helix interactions lead to dimerization of the fused DBD. The dimer is able to bind to the promoter/operator region, controlling reporter gene expression. In the BACTH system, interactions lead to formation of cAMP. cAMP alone is not able to bind to the promoter of the reporter gene and it has to form a complex with cellular CAP first. Thus, the CAP level, which can vary in different cells, might affect reporter gene activity.

6. Finally, it has to be mentioned that all described assays analyze interactions of TM domains within a bacterial membrane. The lipid composition of bacterial membranes, as well as the lipid asymmetry differs substantially, when compared to eukaryotic membranes. As the lipid composition of a membrane can severely influence the propensity of TM domains to form oligomers [52–54], any results obtained by using the bacterial systems have to be interpreted with some caution.

2   Materials

2.1   Molecular Cloning

1. DNA oligomers for molecular cloning are ordered from the manufacturer of choice.

2. 10× in vitro annealing buffer: 200 mM Tris–HCl, pH 7.5, 100 mM $MgCl_2$, 500 mM NaCl, pH 7.5. Filter sterilize.

2.2   Preparation of Competent E. coli Cells and Transformation

Standard molecular biology procedures are described in [55].

1. LB medium and LB agar plates: Dissolve 10 g NaCl, 10 g tryptone, and 5 g yeast extract in 1 L distilled water and autoclave. For preparation of LB agar plates, add 15 g agar per 1 L LB medium prior to autoclaving. After autoclaving, let the agar cool down to approximately 50 °C, add antibiotics at appropriate concentrations, and pour agar plates.

2. Antibiotics: Dissolve the required amount of antibiotic in the indicated solvent (Table 2). Filter sterilize the stock solutions and store at –20 °C. Wrap aliquots of tetracycline in aluminum foil.

3. $CaCl_2$ solution: Dissolve 0.1 M $CaCl_2$ in distilled water and autoclave.

Table 2
Antibiotics stock solutions and final concentrations

Antibiotic	Stock solution	Final concentration (µg/mL)
Ampicillin	100 mg/mL in 50 % ethanol	100
Chloramphenicol	30 mg/mL in ethanol	30
Kanamycin	30 mg/mL in water	30
Tetracycline	10 mg/mL in 50 % ethanol	10
Carbenicillin	100 mg/mL in water	100

Table 3
***E. coli* strains used in ToxR-based assays**

E. coli strains	Relevant features	Reference
FHK12	*F; ara, Δ (lac-proAB), rpsL, thi; φ80 lacZΔM15 ctx::lacZ*	[56]
PD28	*Pop3325 ΔmalE 4444 Δ (srlR-recA)306 ::Tn10*	[57]
NT326	*F-araD139 ΔlacU1169 rpsL thi ΔmalE444 recA1*	[58]

2.3 TOXCAT System and ToxR-Based Systems

1. *E. coli* strains (*see* Table 3).
2. Plasmids (*see* Table 4).
3. TOXCAT sonication buffer: 25 mM Tris–HCl, pH 8.0, 2 mM EDTA.
4. Cell disruption device: Sonifier 250, Branson Ultrasonics.

2.4 GALLEX

1. *E. coli* strains (*see* Table 5).
2. Plasmids (*see* Table 6).

2.5 β-Galactosidase Activity Assay [7]

1. 5× Z-buffer: 300 mM Na_2HPO_4, 200 mM NaH_2PO_4, 50 mM KCl, 5 mM $MgSO_4$.
2. 1× Z-buffer: Dilute 5× Z-buffer 5 times and add β-mercaptoethanol to a final concentration of 50 mM. Always prepare 1× Z-buffer freshly!
3. Dissolve 0.1 % sodium dodecylsulfate (SDS) in water.
4. Dissolve 4 mg/mL *o*-nitrophenyl-β-galactopyranoside (ONPG) in 1× Z-buffer. As ONPG is difficult to dissolve in buffer, eventually crush the non-dissolved pieces with a pipette tip and filter afterwards. Always prepare solution freshly!
5. Dissolve 1 M Na_2CO_3 in distilled water.

Table 4
Plasmids used in the ToxR-based systems

Plasmids	Relevant features	Reference
pToxRI	*toxR* promoter, *cat* (chloramphenicol resistance gene), ColE1 origin of replication	[9]
pToxRIV/ pToxRV	*araBAD* promoter, *araC* (transcriptional regulator for ara operon), *neo* (kanamycin resistance gene), ColE1 origin of replication, myc-tag, His6-tag (not in ToxRIV)	[4]
pToxRVI	*araBAD* promoter, *araC* (transcriptional regulator for ara operon), *neo* (kanamycin resistance gene), ColE1 origin of replication, ToxR S87G mutant	[24]
pToxRVII	*araBAD* promoter, *cat* (chloramphenicol resistance gene), p15A origin of replication, myc-tag, His6-tag	[24]
pToxRVIII	*toxR* promoter, *cat* (chloramphenicol resistance gene), pA15 origin of replication, myc-tag, His6-tag	[24]
pToxRIX	*araBAD* promoter, *araC* (transcriptional regulator for ara operon), *neo* (kanamycin resistance gene), ColE1 origin of replication	[23]
pccKan (*see* **Note 14**)	Derivate of pkk232-8, *toxR* promoter, *bla* (ampicillin resistance gene), *cat* (chloramphenicol resistance gene) fused to *ctx* promoter	[3]

Table 5
***E. coli* strains used in the GALLEX system**

E. coli strains	Relevant features	Reference
SU101	*lexA71::Tn5 (Def)sulA211* Δ*(lacIPOZYA)169/FclacIqlacZDM15::Tn9/sulA op⁺/op⁺::lacZ*	[30]
SU202	*lexA71::Tn5 (Def)sulA211* Δ*(lacIPOZYA)169/FclacIqlacZDM15::Tn9/sulA op408/op⁺::lacZ*	[30]
NT326	*FnaraD139 ΔlacU1169 rpsL thi ΔmalE444 recA1*	[58]

Table 6
Plasmids used in the GALLEX system (*see* Note 15)

Plasmids	Relevant features	Reference
pBLM100 (Km^R)	Derivative of pBR322. Contains P_{tac} promoter and *lacIq* gene from pMal-p2x (New England Biolabs), *bla* (ampicillin resistance gene), *pBM1* origin of replication, *rob*, *neo* (kanamycin resistance gene)	[5]
pALM148 (Km^R)	Derivative of pACYC148. Contains P_{tac} promoter and *lacIq* gene from pMal-p2x (New England Biolabs), *tet* (tetracycline resistance gene), *p15A* origin of replication, *neo* (kanamycin resistance gene)	[5]

Table 7
Plasmids used as controls in the maltose complementation assay

Plasmid	Relevant feature	Reference
pMal-p2: positive control	Contains P_{tac} promoter fused to *malE*, *lacI*^{q} (repressor of P_{tac} promoter), *amp* (ampicillin resistance gene)	New England Biolabs
pMal-c2: negative control	Contains P_{tac} promoter fused to *malE* with deleted signal sequence coding region, *lacI*^{q} (repressor of P_{tac} promoter gene), *amp* (ampicillin resistance gene)	New England Biolabs

2.6 Test for Membrane Insertion of Chimeric Proteins: NaOH Extraction

1. Desoxyribonuclease I from bovine pancreas: Dissolve 10 mg/mL in distilled water.
2. Lysozyme: Dissolve 10 mg/mL in distilled water.
3. Primary antibodies: anti-MalE (e.g., from New England Biolabs). anti-LexA (e.g., from Novus Biologicals).

2.7 Assay for Correct Membrane Insertion: Maltose Complementation Assay

1. The plasmids used as controls are listed in Table 7.
2. M9 agar plates:
   a. Prepare M9 salt solution: Dissolve 12.8 g Na_2HPO_4, 3 g $KH_2PO_4 \cdot 7H_2O$, 0.5 g NaCl, 1 g NH_4Cl, and 15 g agar in distilled water, add water to a final volume of 980 mL and autoclave.
   b. Prepare separate solutions of 1 M $MgSO_4$ and 1 M $CaCl_2$ and sterilize by autoclaving.
   c. Prepare fresh solutions of 20 % maltose and 20 % glucose (w/v) in distilled water separately and filter sterilize.
   d. Let the autoclaved M9 salts/agar solution cool down to about 50 °C after autoclaving and add 2 mL 1 M $MgSO_4$, 20 mL of the appropriate carbon source (20 % glucose or 20 % maltose solution) and 0.1 mL 1 M $CaCl_2$.
   e. Cast into sterile petri dishes.
3. M9 liquid media:
   a. Prepare as described above (M9 agar plates **steps a–d**) without adding agar.
   b. Phosphate-buffered saline (PBS): Dissolve 137 mM NaCl, 2.7 mM KCl, 10 mM Na_2HPO_4, and 2 mM KH_2PO_4 in distilled water, adjust the pH to 7.4 and autoclave afterwards.

2.8 Nonradioactive CAT Assay

1. CAT-assay reaction buffer (*see* **Notes 16** and **17**):
   a. Tris buffer: 1 M Tris–HCl, pH 7.8.
   b. Dissolve 4 mg 5,5′-dithiobis-(2-nitrobenzoic acid) (DTNB) in 1 mL 1 M Tris–HCl buffer.

c. Add 0.2 mL of a fresh 5 mM acetyl coenzyme A solution (trilithium salt). Add molecular biology grade water to a final volume of 10 mL.

2. Chloramphenicol solution: 5 mM chloramphenicol in ethanol.

3 Methods

3.1 *In Vitro* Annealing

To generate a double-stranded DNA fragment coding for a TM helix sequence, two complementary, single-stranded DNA oligonucleotides might be annealed in vitro to generate a double-stranded DNA cassette for ligation.

1. To 43 μL of molecular biology grade water, add 5 μL of tenfold concentrated in vitro annealing buffer and 1 μL of each of the two corresponding DNA oligomers at a respective concentration of 10 pmol/μL.
2. Heat to 95 °C for 10 min.
3. Let the reaction tube slowly cool down to room temperature (should take about 30 min).
4. Incubate on ice for additional 10 min.

3.2 Molecular Cloning

Prior to ligation of the generated DNA cassettes into the plasmid(s) of the respective system, restriction digestion of the plasmid(s) is necessary. Figure 5 shows the cloning sites of the plasmids needed for the TOXCAT- and GALLEX system. This is performed following the protocols provided by the (enzyme) manufacturers.

To extract the restriction-digested plasmid from an agarose gel, a commercially available extraction kit may be used. For ligating the generated DNA cassettes into the respective plasmid, the T4 DNA ligase can be used according to the manufacturers' instructions. The plasmid-to-insert ratio was at least 1:5 in a total volume of 20 μL.

3.3 Preparation of Transformation-Competent *E. coli* Cells

1. Pick a single *E. coli* colony from an LB plate and inoculate an overnight culture of 5 mL LB. Add antibiotics (Table 2) according to the *E. coli* strain used and/or already transformed plasmids.

a

```
...ToxR                              NheI              BamHI                          MalE...
cca aac ttg ggg aat cga GCT AGC gct ... cgg aGG ATC Ctg atc aac cca agc caa tcc
P   N   L   G   N   R           ..........  R   I   L   I   N   P   S   Q   S
```

b

```
...LexA                         NheI        SacI                                       SpeI        BamH I       MalE...
...gtgtggct gcc ggt gaa cca GCT AGC GGG AGC TCG ctt cac gct ....gagcggg act ctg ACT AGT AGG ATC CTG atc aac cca
            G   E   P   A   S   G   S   S   L   H   ..........                 T   L   T   S   R   I   L   I   N
```

Fig. 5 Multiple cloning sites of the plasmids *pccKan* (**a**) used in the TOXCAT system and the sequences of the *pALM* and *pBLM* plasmids (**b**) used in the GALLEX system

2. The next day, give 0.5 mL of the overnight culture to 100 mL LB medium (antibiotic added) and grow the cells until an OD_{600} of about 0.6 is reached.

3. Place the cells on ice for 10 min and pellet them afterwards by centrifugation (4,000×g, 10 min, 4 °C).

4. Discard the supernatant, resuspend the cell pellet in 10 mL ice-cold 0.1 M $CaCl_2$ solution and incubate on ice for 10 min.

5. Pellet cells again by centrifugation, discard supernatant, and resuspend cells in 2 mL ice-cold 0.1 M $CaCl_2$ solution.

6. For the transformation, add 0.1 μg plasmid to 200 μL of competent cells and incubate on ice for 20 min.

7. Heat shock the cells for 2 min at 42 °C.

8. Immediately chill the cells on ice for 10 min.

9. Add 800 mL LB medium and incubate at 37 °C in a rotary incubator for 1 h.

10. Streak out 200 μL of the culture onto LB agar plates, containing the appropriate antibiotics and incubate the plates overnight at 37 °C.

11. For GALLEX measurements two plasmids have to be transformed, as described in Subheading 3.4.3.

3.4 Transformation and E. coli Culture

3.4.1 ToxR-Based Systems

Depending on the ToxR system, plasmids are used in different combinations. For measuring homotypic interactions, pToxRI, pToxRIV or pToxRV have been developed [2, 4, 10, 56, 59]. The differences of the plasmids and the recommendations for their application are mentioned in **Note 4**. The positive ToxR system has been developed for assaying heterotypic interactions of two helices whose homodimer formation is significantly weaker than the heterodimer formation [23]. The strength of heterodimer formation is evaluated by comparing the mean β-galactosidase activity of *E. coli* cells transformed with only one helix to the mean β-galactosidase activity of *E. coli* cells transformed with both helices. In order to measure homodimerization, each helix has to be expressed separately from the plasmids pToxRIX and pToxRVII, and for measuring heterodimerization, one helix has to be expressed from pToxRIX and the second helix from pToxRVII and vice versa. The plasmids differ in their per cell copy number; pToxRIX is a low-copy-number plasmid and pToxRVII is a high-copy-number plasmid. To be able to distinguish between the proteins derived from the high- and the low-copy-number plasmid on an acrylamide gel, the his6-tag and the myc-tag of pToxRIX have been removed. For measuring heterotypic interactions with the dominant-negative system, the plasmids pToxRVI, pToxRVII, and pToxRVIII are necessary [60]. It has been developed to investigate whether homotypic interactions compete for heterotypic interactions with another helix. The helix forming a homodimer is either expressed from pToxRVII or pToxRVIII encoding the wt ToxR DBD and differing in their promoters, whereas the competing helix is expressed from pToxRVI encoding the inactive ToxR S87G DBD (*see* **Note 6**). Heterodimer formation results in reduced β-galactosidase activity compared to homodimerization. Table 4 summarizes plasmids needed for ToxR-based systems.

ToxR Assay for Measuring Homotypic Interactions

1. Prepare competent *E. coli* FHK12 cells as described above.

2. Transform ToxR plasmids (pToxRI: constitutive protein expression, pToxRIV and pToxRV: arabinose-inducible protein expression) and streak the cells out on agar plates containing 100 μg/mL ampicillin (FHK12 cells are ampicillin-resistant) and a second antibiotic depending on the transformed plasmid. Incubate agar plates at 37 °C overnight.

3. On the next day pick three single colonies to inoculate 3 mL LB medium and incubate at 37 °C and 200 rpm. Depending on the transformed plasmid, the composition of the medium might differ:

   pToxRI: 100 μg/mL ampicillin, 30 μg/mL chloramphenicol, 2 % maltose, 0.4 mM IPTG (optional). IPTG enhances the dynamic range of β-galactosidase activity, so that subtle differences in TM helix–helix interactions between related sequences might be determined better.

   pToxR IV and pToxRV: 100 μg/mL ampicillin, 33 μg/mL kanamycin, 0.4 mM IPTG (optional).

4. Take 250 μL of the overnight culture and inoculate 10 mL LB medium. Again the antibiotics added to the LB medium might differ (Subheading 3.4.1.1, **step 3**). Add up to 0.01 % arabinose to the FHK12 cells transformed with the plasmids ToxRIV and ToxRV to induce protein expression (*see* **Note 18**).

5. Incubate cell culture at 37 °C and 200 rpm in a rotary incubator.

6. When cells have reached an OD_{600} of about 0.6, start β-galactosidase activity measurements (Subheading 3.5).

7. Analyze protein expression levels (Subheading 3.7) and the topology of the chimeric proteins (Subheading 3.8).

Dominant-Negative or Positive ToxR Assay for Measuring Heterotypic Interactions

1. Prepare competent *E. coli* FHK12 cells.

2. Transform competent cells either with pToxRVI (dominant-negative assay) or pToxRIX (positive assay) and streak the cells out on LB agar plates containing 33 μg/mL kanamycin. Incubate cells overnight at 37 °C.

3. The next day, pick a single colony and prepare competent cells as described.

4. Transform pToxRVI-carrying competent cells with pToxRVII or pToxRVIII (dominant-negative system), respectively, and the pToxRIX-carrying cells with pToxRVII (positive system). Streak cells out on LB agar plates containing 33 µg/mL kanamycin and 30 µg/mL chloramphenicol. Incubate overnight at 37 °C.

5. On the next day, pick three single colonies and inoculate 3 mL LB medium containing 100 µg/mL ampicillin, 33 µg/mL kanamycin, and 30 µg/mL chloramphenicol. Incubate overnight on a rotary incubator at 37 °C and 200 rpm.

6. On the next day inoculate 10 mL fresh LB medium containing 100 µg/mL ampicillin, 33 µg/mL kanamycin, 30 µg/mL chloramphenicol, and up to 0.01 % arabinose (*see* **Note 19**) with 250 µL of the overnight cultures. The amount of arabinose has to be adjusted for each experiment (*see* **Note 18**). Perform a β-galactosidase assay when the cells have reached an OD_{600} of approximately 0.6 (Subheading 3.5).

7. Check the amount of protein inserted into a membrane by Western blot analysis as well as the membrane topology of the expressed proteins (Subheadings 3.7 and 3.8).

3.4.2 TOXCAT

Since the TOXCAT system measures homooligomerization of TM helices, only one plasmid coding for the chimeric protein containing a TM domain of interest needs to be transformed into *E. coli* cells. Expression of the TM domain-containing protein is controlled by the constitutive *toxR* promoter; thus, no induction of protein expression is required.

1. Transform the generated plasmid into *E. coli* NT326 cells. Additionally transform plasmids encoding the TM helices of human wt GpA (strong interaction) and its G83I mutant (weak interaction) [5, 61] as controls (*see* **Note 2**).

2. Plate the transformed cells on LB agar plates containing 100 µg/mL ampicillin.

3. Pick a minimum of three colonies per variant that is to be measured from each plate and transfer into 2.5 mL liquid LB medium containing carbenicillin.

4. Incubate overnight at 37 °C in a rotary incubator.

5. Dilute the overnight culture 1:50 in LB medium containing carbenicillin by adding 400 µL of the overnight culture to 20 mL LB medium in an Erlenmeyer flask.

6. Grow culture to an OD_{600} of 0.6.

7. Aliquot 1 mL of the culture into a 2 mL reaction tube. If the culture is above or below the required optical density, adjust the volume to obtain an equal amount of cells. Take three aliquots, respectively, to perform the later measurements in triplicates. If desired also take 1 mL culture of an $OD_{600nm} = 0.6$ and use for NaOH extraction.

From this step on, keep cell suspensions, pellets, or lysates on ice.

8. Centrifuge cells for at least 1 min at 16,000×g at 4 °C. Remove supernatant.

9. Resuspend the cell pellet in 1 mL sonication buffer.

10. Centrifuge again for at least 1 min at 16,000×g at 4 °C. Discard the supernatant.

11. Resuspend the cell pellet in 2 mL sonication buffer.

12. Lyse cell suspension by ultrasonication.

13. Centrifuge lysate for 15 min, 16,000×g at 4 °C, and transfer supernatant to a new reaction tube (Note: At this point, the lysate can be quick-frozen in liquid nitrogen and stored at –70 °C).

3.4.3 GALLEX

Depending on the type of interaction to be measured, the adequate plasmids and *E. coli* strains have to be chosen. Homotypic interactions can be measured in the *E. coli* reporter strain SU101 (Table 5) together with the pBLM100 plasmid (Table 6). For measuring heterotypic interactions, the reporter strain SU202 (Table 5) is used in combination with the plasmids pBLM100 and pALM148 (Table 6). The plasmids differ in the encoded LexA DBD. On pBLM100 the wt LexA DBD is encoded whereas pALM148 encodes the mutated LexA DBD.

Homooligomerization of TM Domains

1. The plasmid pBLM100 is used together with the *E. coli* strain SU101 to measure TM helix homooligomerization (*see* **Note 20**). A DNA cassette encoding the TM helix of interest is ligated to the plasmid pBLM100 using two suitable restriction sites (Fig. 5).

2. Prepare competent SU101 cells as described (*see* **Note 21**).

3. Transform competent SU101 cell by heat shock with the respective pBLM100-based plasmid. Streak out 200 µL of the cell culture onto agar plates containing 100 µg/mL ampicillin and incubate overnight at 37 °C.

4. The next day, pick at least three independent colonies from each transformation and inoculate 3 mL LB media supplemented with 100 µg/mL ampicillin, 5 µg/mL chloramphenicol, 5 µg/mL kanamycin, and 0.01 mM IPTG. Incubate overnight at 37 °C on a rotary incubator at 200 rpm.

5. Take 250 µL of the overnight cultures and inoculate 10 mL fresh LB medium supplemented with the antibiotics and IPTG in a 50 mL Erlenmeyer flask (*see* **Note 22**). When an OD$_{600}$ = 0.6 is reached, perform the β-galactosidase activity assay.

6. Proper insertion of the expressed chimeric proteins into the *E. coli* inner membrane has to be analyzed (Subheading 3.8), as well as the relative amount of membrane-inserted proteins (Subheading 3.7).

Heterooligomerization of TM Domains

1. For analyzing interactions of two different TM helices, the *E. coli* reporter strain SU202 has to be used together with the plasmids pALM148 and pBLM100. A DNA cassette encoding the TM helix of interest has to be ligated to pBLM100, and the TM helix is thereby fused to the wt LexA DBD. A DNA cassette encoding another TM helix is genetically fused to the mutated DBD of LexA by ligating its coding sequence to pALM148. The multiple cloning site of pALM148 is identical to the MCS of pBLM100 (Fig. 5) and ligation is performed as described (Subheading 3.2).

2. Prepare competent *E. coli* SU202 cells as described above.

3. Transform SU202 first with the pALM148-derived plasmids (*see* **Note 23**). Streak the cells out on LB agar plates containing 10 μg/mL tetracycline and 5 μg/mL chloramphenicol. Incubate at 37 °C overnight.

4. Pick a single colony of each transformant the next day and inoculate 3 mL LB medium, containing 10 μg/mL tetracycline. Incubate overnight at 37 °C and 200 rpm in a rotary incubator.

5. Inoculate 50 mL LB medium containing 10 μg/mL tetracycline with 250 μL of the overnight culture. Prepare competent cells when cells have reached an O_{600} of ~0.6.

6. Transform the pBLM100-derived plasmids. Streak cells out on LB agar plates containing 100 μg/mL ampicillin and 10 μg/mL tetracycline. Incubate overnight at 37 °C.

7. The next day pick at least three single colonies from each transformation and inoculate 2 mL LB medium with 100 μg/mL ampicillin, 10 μg/mL tetracycline, 5 μg/mL chloramphenicol, 5 μg/mL kanamycin, and 0.01 mM IPTG. Incubate overnight at 37 °C and 200 rpm in a rotary incubator.

8. Take 10 mL fresh LB medium with antibiotics and add 250 μL of the overnight culture. Adjust the IPTG concentration as described (*see* **Note 22**). Let the cells grow in a 50 mL Erlenmeyer flask until an $OD_{600} = 0.6$ is reached and perform β-galactosidase activity measurements.

9. Check the orientation of the fusion proteins by the maltose complementation assay (Subheading 3.8) and confirm equal expression of the individual chimeric proteins (Subheading 3.7).

3.5 β-Galactosidase Activity Assay

1. Let an *E. coli* culture grow until the cell density has reached an OD_{600} of approximately 0.6 and record the exact cell density of the culture.

2. Take 100 μL of the cells and add 900 μL of 1× Z-buffer (β-mercaptoethanol added).

3. Add 10 µL of 0.1 % SDS, 10 µL chloroform, and vortex for 10 min to lyse the cells.

4. Start the reaction by adding 200 µL ONPG (4 mg/mL in 1× Z-buffer) to each tube and mix by inverting the tube twice. Add the ONPG solution from sample to sample in 10 s increments to the next sample. Ten seconds turned out to be a suitable time period to comfortably pipette ONPG in one tube, mix, and pipette ONPG into the next tube.

5. Incubate at room temperature and record the time it took for the yellow color to appear. As the negative control (empty expression plasmid) is supposed to show the most intense yellow color (in the GALLEX system), start with the addition of ONPG to the negative control. In the ToxR system the strongly interacting GpA TM helix might be used as a positive control and be placed first.

6. When the sample becomes noticeably yellow, stop all the reactions. However, the reaction should run at least for 3 min. If the reaction is completed sooner, repeat with only 50 µL of the cells.

7. Stop the reaction by adding 500 µL of 1 M NaCO$_3$ to each tube and mix by inverting the tube twice. Continue with the next tube after 10 s to make sure that all reactions have been run for exactly the same time.

8. Centrifuge the samples for 1 min at maximum speed.

9. Transfer the supernatant to a cuvette and record absorbance at 550 and 420 nm.

10. Calculate the β-galactosidase activity in Miller units [7].

$$(1{,}000 \times \mathrm{OD}_{420} - (1.75 \times \mathrm{OD}_{550}))/\text{incubation time (min)} \times \text{vol}$$
$$(0.1\ \mathrm{mL}) \times \mathrm{OD}_{600}$$

3.6 Nonradioactive CAT Assay

The CAT activity in cell lysates can be measured nonradioactively by a spectrophotometric assay [62], as CAT catalyzes the transfer of an acetyl group from acetyl coenzyme A to chloramphenicol, releasing coenzyme A. The free thiol group of coenzyme A will react with 5,5-dithiobis-(2-nitrobenzoic acid) (DTNB), leading to a color change that can be quantified spectrophotometrically.

1. Use the CAT-assay reaction buffer as reference for the photometric measurements. The absorption will be measured at 412 nm for at least 2 min in intervals of 5 s or less.

2. Add 20 µL cell lysate (*see* Subheading 3.4.2, **step 12**) to 500 µL of the CAT-assay reaction buffer in a cuvette and place sample into the spectrophotometer (*see* **Notes 16** and **17**).

3. Monitor the acetyl coenzyme A hydrolysis in the CAT-assay reaction buffer for 2 min.

4. After 2 min, add 10 μL of 5 mM chloramphenicol solution (*see* **Note 24**).

5. Follow OD_{412} for additional 2 min.

6. Generate regression lines for the measured values to account for any change in absorption caused by acetyl coenzyme A background hydrolysis (first 2 min) and for the values measured after addition of chloramphenicol (*see* **Note 25**).

7. Calculate the slopes of both regression lines.

8. To determine the enzymatic activity, use the following equation:

$$\text{Enzymatic activity} = (y - x) \times 13.6^{-1}$$

where the DTNB extinction coefficient $\varepsilon = 13.6$ [62], x is the slope of the regression line of measured values of acetyl coenzyme A background hydrolysis, and y the slope of the regression line of values measured after chloramphenicol addition.

3.7 Test for Membrane Insertion of Chimeric Proteins: NaOH Extraction

The relative concentration of chimeric proteins expressed into the *E. coli* membrane needs to be verified for each analyzed protein. Therefore, *E. coli* cells are treated with lysozyme after harvesting, and soluble proteins are extracted with NaOH [5, 38]. Subcellular localization of the chimeric protein is detected by Western blot analysis. A primary anti-MalE antibody might be used for detection of fusion proteins expressed in the GALLEX and ToxR-based systems. For detection of chimeric protein expressed in the GALLEX system, a commercially available anti-LexA antibody is also suitable.

1. Grow cell culture as described for the respective system (TOXCAT (Subheading 3.4.2), GALLEX (Subheading 3.4.3)) until culture has reached an OD_{600} of ~0.6. Take an equivalent of 1 mL of an $OD_{600nm} = 0.6$ and pellet by centrifugation.

2. Resuspend cell pellet in 90 μL molecular biology grade water, 2.5 μL 1 M $MgCl_2$, 5 μL DNase (10 mg/mL), and 5 μL lysozyme (10 mg/mL).

3. Incubate at room temperature for 1 h. Place on ice afterwards.

4. Add 150 μL ice-cold, molecular biology grade water and mix.

5. Aliquot 125 μL and mark as "whole cell lysate." Store on ice.

6. Add 125 μL ice-cold 0.1 M NaOH to the remaining lysate.

7. Mix rigorously for 1 min with a vortex device.

8. Centrifuge for 15 min at $16,000 \times g$ and 4 °C.

9. Transfer supernatant into another reaction tube. Keep the pellet on ice and mark as "membrane proteins."

10. Centrifuge the supernatant in an ultracentrifuge for 30 min at $40,000 \times g$, 4 °C.

11. Transfer supernatant to another reaction tube, mark as "soluble proteins," and keep on ice.

12. Precipitate proteins of the "whole cell lysate" and the "soluble proteins" fraction by adding trichloroacetic acid (50 %, w/v) at a volume half of the protein solutions' volumes.

13. Incubate on ice for 30 min.

14. Centrifuge at $16,000 \times g$ for 15 min at 4 °C and remove supernatant completely using a pipette. Discard supernatant.

15. Add 1 mL ice-cold acetone to pellet and incubate on ice for 5 min.

16. Centrifuge at $16,000 \times g$ for 15 min at 4 °C and remove supernatant afterwards completely using a pipette.

17. Air-dry the pellets under the fume-hood.

18. Add 50 μL of onefold SDS sample buffer to the air-dried pellets and the "membrane proteins" pellet (*see* **Note 26**).

19. Perform SDS-PAGE and subsequent Western blot analysis following standard procedures.

3.8 Control for Correct Membrane Insertion: Maltose Complementation Assay

Due to the different properties of individual TM helices in the expressed chimeric proteins, e.g., caused by the varying amino acid sequences, proper orientation and insertion into the inner *E. coli* membrane has to be analyzed for every expressed chimeric protein. Correct membrane insertion and the TM topology might be analyzed by a maltose complementation assay [19], which assays the activity of the MalE domain in the *E. coli* periplasm. When the chimeric protein is expressed in a MalE-deficient strain, growth is only possible on agar plates or in liquid medium containing maltose as the only carbon source, when maltose uptake from the medium into the cell is mediated by the periplasmic MalE domain (*see* **Note 27**). Growth on minimal medium agar plates containing maltose as the only carbon source will confirm that maltose uptake into the cell is mediated by the protein's MalE domain being present in the *E. coli* periplasm. Addition of glucose instead of maltose serves as a positive control for monitoring the viability of the cells. As a further control, pMal-p2 and pMal-c2 vectors are also transformed into the *malE*-deficient *E. coli* strain. From the pMal-p2 vector MalE is expressed with the N-terminal signal sequence, and thus the protein will be transported to the periplasmic space (positive control). The *malE* gene in the pMal-c2 vector lacks the signal sequence coding region. Thus, the encoded protein remains in the *E. coli* cytosol after expression and maltose cannot be transported into the cell (negative control).

3.8.1 Growth on Minimal Medium Agar Plates

1. Prepare M9 agar plates containing 0.4 % (w/v) maltose and M9 agar plates containing 0.4 % (w/v) glucose. If required (*see* **Note 28**), add IPTG to a final molar concentration of 500 μM.

2. Transform the respective plasmids into *E. coli* NT326 cells. As controls, additionally transform the pMal-p2 and pMal-c2 plasmids.

3. Plate cells after transformation on LB agar plates with appropriate antibiotics. Incubate at 37 °C overnight.

4. Use an inoculation loop to transfer a single colony from the LB agar plate onto M9 plates, containing maltose or glucose, respectively.

5. Incubate plates for 2–4 days at 37 °C.

3.8.2 Growth in Liquid Minimal Medium

Besides analyzing growth of *malE*-deficient *E. coli* strains on minimal medium agar plates, growth can also be monitored in liquid minimal medium to test for the correct topology of an expressed chimeric protein. This is typically done when the ToxR assay is applied [23, 60, 63].

1. Transform *E. coli malE*-deficient cell (e.g., PD28 cells) with the plasmids used for the ToxR assay and plate the cells on LB agar plates with the appropriate antibiotics. Incubate overnight at 37 °C.

2. Inoculate 5 mL LB medium supplemented with the appropriate antibiotics with a single colony and incubate overnight at 37 °C in a rotary incubator at 200 rpm.

3. Pellet cells by centrifugation ($3,000 \times g$, 10 min, 4 °C).

4. Discard supernatant and wash the cells twice in 2 mL PBS. Resuspend the pellet in 2 mL PBS and pellet cells by centrifugation as described. Repeat this step.

5. Finally, resuspend the cells in 2.5 mL PBS. Take 300 μL of the cell suspension and inoculate 50 mL minimal medium in an Erlenmeyer flask. Let the cells grow for 20–24 h at 37 °C. Follow cell growth by recording OD_{600} in appropriate intervals for up to 50 h.

4 Notes

1. In all variants of the ToxR-based systems and in the GALLEX system, a single α-helix is fused to the soluble MalE protein and a soluble DBD. These domains are significantly larger than the TM helix, and thus the fusion domains may influence the interactions due to spatial problems. It has been reported that, for example, the TOXCAT and the ToxR system are

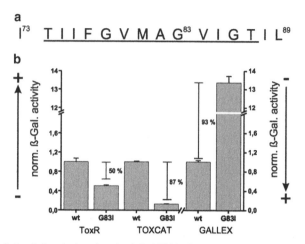

Fig. 6 β-Gal activity of wt and mutant GpA TM helix obtained via the classical ToxR system, the TOXCAT system and the GALLEX system, normalized to wt GpA. (**a**) The amino acid sequence of the wt GpA TM region used in the GALLEX system consists of 17 amino acids. The sequence best suitable for ToxR measurements is only 13 amino acids long (*underlined*). (**b**) The wt GpA helix displays a strong interaction propensity, whereas the GpA G83I mutant has only a reduced interaction propensity. In the ToxR and the TOXCAT system, where helix–helix interaction activates reporter gene expression, the β-Gal activities measured for the G83I mutant are lower than those measured for the GpA wt. In the GALLEX system, the reporter gene is inactivated after binding of the LexA dimer. Thus, the β-Gal activity measured for GpA G83I is higher than in the case of the wt GpA TM helix. The black arrows indicate the corresponding interaction propensity

sensitive to the length of the fused TM helix [2]. Nevertheless, the fusion to the MalE protein has indispensable advantages. MalE ensures a proper topology of the expressed chimeric protein, facilitates insertion into the membrane, and offers the possibility to assess both properties in a maltose complementation assay (Subheading 3.8) [3, 9].

2. It is recommended to use positive as well as negative controls in each measurement. The TM domain of glycophorin A (GpA) interacts strongly and the G83I mutant of GpA weakly. Both helices have been successfully used in GALLEX measurements [5]. In Fig. 6 the amino acid sequence of the GpA TM helix is given. This sequence has turned out to have strong interaction propensities in ToxR/TOXCAT [3] and GALLEX measurements [5].

3. Noteworthy, when assayed with any ToxR-based or the GALLEX system, the N-terminal end of the GpA TM domain is located in the bacterial cytoplasma, whereas its original

orientation in an erythrocyte cell membrane is vice versa [64]. Whether this has an effect on the measurements still needs to be shown.

4. In the early developed ToxRI plasmids, expression of the chimeric proteins is constitutive, whereas the later ToxRIV and ToxRV plasmids offer the possibility to tightly control protein expression, as expression is controlled by the inducible *ara* promoter. By varying the inducer concentration (arabinose or arabinose and glucose) the amount of expressed and membrane-integrated protein can be regulated. Since the measured propensity of a TM helix to oligomerize depends on its concentration within the membrane, different expression levels might have to be tested to identify optimal expression conditions. The difference in reporter gene activity between the negative and the positive control should be significant. Therefore, the use of the inducible promoters is recommended as it allows fine-tuning of experimental conditions.

5. In contrast to the POSSYCCAT system, in the TOXCAT system the *ctx::cat* sequence is located on a multicopy plasmid. In the assay multiple ToxR chimeric proteins will be present in the *E. coli* inner membrane, and thus it is possible that the presence of several ToxR dimer-binding promoters offers a higher possibility to analyze differences in homooligomerization propensities. Furthermore, possible interference by a genomic DNA structure is avoided by using a plasmid-encoded reporter gene [65]. On the other hand, it has been suggested that by having only a single copy of the *cat* reporter gene in the *E. coli* genome, discrimination of different interactions propensities is favored [4].

6. The wt ToxR domain is expressed from a low-copy plasmid containing either the constitutive *toxR* (ToxRVIII) or the inducible *ara* (ToxRVII) promoter, whereas the inactive ToxR S87G domain is expressed from a high-copy plasmid, containing the inducible *ara* promoter. In principle, a reduced reporter gene activity could simply be caused by diminished wt ToxR expression. To exclude this possibility, equal expression levels of both chimeric proteins have to be confirmed by Western blot analysis (Subheading 3.7) [24, 60]. Constitutive expression of the wt ToxR domain offers the possibility to gradually increase the ToxR S87G expression by adding arabinose and therefore to adjust the amount of expressed ToxR S87G to the wt ToxR amount.

7. A further limitation in heterodimerization measurement with the GALLEX system is the use of two different plasmids. The helix, that is C-terminally fused to the wt LexA DBD, is expressed from pBLM100, whereas the second TM helix is genetically fused to LexA148 and expressed from the

pALM148 plasmid. Although both plasmids are low-copy-number plasmids, they still differ in their cellular copy number, resulting in different levels of the expressed chimeric proteins. Thus, expression of two different TM helices from two different plasmids might result in different interaction propensities [5, 31], and an interaction propensity always has to be measured in both orientations, i.e., each TM helix has to be expressed in independent experiments from pBLM100 as well as from pALM148, respectively, together with the second helix expressed from the remaining plasmid.

8. As the strength of reporter gene repression depends on the amount of membrane-integrated, chimeric protein, the appropriate IPTG concentration for the actual experiment might have to be determined. When interactions of the wt GpA or G83I-mutated GpA TM helices have been monitored at IPTG concentrations ranging from 0.005 to 0.1 mM, a maximum of 50–80 % difference has been observed at medium IPTG concentrations [5]. While addition of increasing IPTG concentration results in stronger *lacZ* repression (due to an increased concentration of the expressed protein within the membrane), at high IPTG concentration strongly interacting TM helices might not be properly distinguished anymore from more weakly interacting ones [31].

9. Despite the advantages of the GALLEX system, it also has its limitation: It has to be considered that the two large soluble domains fused to a single TM helix may influence interaction propensities of the TM helix, as discussed already for the ToxR-based systems (*see* **Note 1**). It has also been shown that the GALLEX system is sensitive to the orientation of an interacting TM helix surface relative to the soluble domains. In order to find the proper orientation, helices of different lengths have to be used and the sequence has to be rotated relative to the fusion domains [5]. Although time consuming and eventually even impossible (especially in case of heterooligomerization), evaluation of possible interactions of different sequences encoding the same TM helix (but varying in length and orientation to the fused domains, etc.) might be necessary to be able to draw any final conclusion [5].

10. The TOXCAT/ToxR- and GALLEX system, but not the BACTH system, share a basic functional property: A dimeric DBD, that is part of an integral membrane protein, binds to a promoter/operator region localized in the *E. coli* genome or on a plasmid. Thus, a reporter gene expression directly correlates with the amount of DNA-bound, dimerized DBD.

11. The strength of a reporter gene expression thus depends on the cAMP/CAP level in the cell, whereas binding of DNA to cytosolic binding domains of integral membrane proteins requires the DNA to be present near the membrane. This could lead to

a stronger signal in the cAMP-based system compared to the systems using fusion proteins containing DBDs.

12. In the BACTH system, the two generated plasmids have significantly different copy numbers and one plasmid is a low-copy-number plasmid whereas the other has a (moderately) high copy number. Thus, the resulting differences in the amount of expressed proteins will certainly influence the measured interaction propensities, and thus any results have to be interpreted with great care.

13. In GALLEX system and ToxR-based systems, TM helices can only be incorporated into the inner membrane with their respective N-terminal ends facing the cytosol (parallel orientation).

14. The insertion of the fragment can be checked by growth on kanamycin-containing agar plates. The plasmid *pccKan* will lose its kanamycin resistance when the plasmid is restriction digested by *BamHI* and *NheI*. Thus, colonies of transformed cells containing the TM helix coding region transferred on kanamycin agar plates will not grow. However, correct insertion should always be confirmed by DNA sequencing. The inserted fragment must match the reading frame of the ToxR-MalE-chimeric construct. Use common *E. coli* strains for molecular cloning, e.g., XL1-blue or DH5α.

15. For cloning reasons a *neo* cassette has originally been introduced in the multiple cloning sides of both pBLM100 and pALM148. As after restriction digestion of the respective plasmids the overall size of the plasmids is reduced by ~1,200 bp, successful restriction digestion can be conveniently monitored on agarose gels.

16. 10 mL CAT-assay reaction buffer is sufficient for 18 measurements plus one blank sample.

17. Prepare CAT-assay reaction buffer immediately before starting the measurements.

18. The appropriate arabinose concentration has to be established individually for each assay. It is advisable to take the arabinose concentration where the protein expression level leads to a pronounced difference in β-galactosidase activities between a positive and a negative control.

19. For the dominant-negative assay, it is suggested to grow cells in the presence of 0.4 mM IPTG, as IPTG might enhance the differences between signals elicited by TM domains of different affinity [2].

20. While in the GALLEX system two plasmids are transformed into *E. coli* SU202 for analyses of heterotypic interactions, only one plasmid is transformed here and the topology of only

a single expressed chimeric protein is assayed at a time, as the assay does not allow discriminating wether only a single or both expressed proteins are properly membrane integrated.

21. Use, e.g., 30 μg/mL chloramphenicol in the overnight culture and 30 μg/mL kanamycin for the culture on the next day. This can serve as a control to verify that the correct *E. coli* strains are used. The SU101 and SU202 strains contain chromosomally integrated chloramphenicol and kanamycin resistance genes.

22. The IPTG concentration in the main culture is crucial, as the amount of expressed and membrane-integrated protein varies depending on the IPTG concentration. As the interaction propensity of the expressed proteins depends on the concentration of the chimeric proteins integrated into the membrane, the IPTG concentration might have to be adjusted from case to case to achieve optimal results. At 0.5 mM IPTG the promoter is fully induced.

23. It has turned out that the pALM-derived plasmid is more stable in the *E. coli* strain than the pBLM-derived plasmid. Therefore, the pALM-based plasmid should be transformed first.

24. If possible, leave the cuvette in the photometer and mix by pipetting up and down.

25. For generation of regression lines, make sure to leave out the values measured during addition of the chloramphenicol solution into the cuvette.

26. If the pellet does not dissolve properly, add increased volumes of SDS sample buffer. If the pH of the resuspended pellets is too low—as indicated by a yellow color—adjust by adding NaOH solution until color changes to blue.

27. While this assay confirms that the MalE domain is localized within the *E. coli* periplasm after expression, it still is possible that a fraction of the expressed protein has an inverse topology, i.e., the chimeric protein inserts into the *E. coli* membrane with a dual topology. A more direct proof of the topology would, e.g., be an assay using spheroplasts [3]. Addition of proteases to stable spheroplasts results in proteolysis of solely the periplasmic protein domains. Protection of cytoplasmic and proteolysis of periplasmic domains can subsequently be analyzed by Western blot analyses using appropriate antibodies. However, thus far a dual topology has never been described and as the spheroplast assays are experimentally highly demanding, the maltose complementation assay became the accepted standard topology prediction assay.

28. When chimeric proteins are expressed from the *tac* promoter (GALLEX system), add isopropyl β-D-1-thiogalactopyranoside (IPTG) to a final molar concentration of 500 μM.

Acknowledgements

This work was supported by grants from the Stiftung Rheinland-Pfalz für Innovation, the Deutsche Forschungsgemeinschaft, the Research Center "Complex Materials" (COMATT), and the University of Mainz.

References

1. Bowie JU (2005) Solving the membrane protein folding problem. Nature 438(7068): 581–589

2. Langosch D et al (1996) Dimerisation of the glycophorin A transmembrane segment in membranes probed with the ToxR transcription activator. J Mol Biol 263(4):525–530

3. Russ WP, Engelman DM (1999) TOXCAT: a measure of transmembrane helix association in a biological membrane. Proc Natl Acad Sci USA 96(3):863–868

4. Gurezka R, Langosch D (2001) In vitro selection of membrane-spanning leucine zipper protein-protein interaction motifs using POSSYCCAT. J Biol Chem 276(49): 45580–45587

5. Schneider D, Engelman DM (2003) GALLEX, a measurement of heterologous association of transmembrane helices in a biological membrane. J Biol Chem 278(5):3105–3111

6. DiRita VJ, Mekalanos JJ (1991) Periplasmic interaction between two membrane regulatory proteins, ToxR and ToxS, results in signal transduction and transcriptional activation. Cell 64(1):29–37

7. Miller VL, Mekalanos JJ (1984) Synthesis of cholera toxin is positively regulated at the transcriptional level by toxR. Proc Natl Acad Sci USA 81(11):3471–3475

8. Miller VL, Taylor RK, Mekalanos JJ (1987) Cholera toxin transcriptional activator toxR is a transmembrane DNA binding protein. Cell 48(2):271–279

9. Kolmar H et al (1995) Membrane insertion of the bacterial signal transduction protein ToxR and requirements of transcription activation studied by modular replacement of different protein substructures. EMBO J 14(16): 3895–3904

10. Brosig B, Langosch D (1998) The dimerization motif of the glycophorin A transmembrane segment in membranes: importance of glycine residues. Protein Sci 7(4):1052–1056

11. Zhou FX et al (2001) Polar residues drive association of polyleucine transmembrane helices. Proc Natl Acad Sci USA 98(5):2250–2255

12. Kubatzky KF et al (2001) Self assembly of the transmembrane domain promotes signal transduction through the erythropoietin receptor. Curr Biol 11(2):110–115

13. Mendrola JM et al (2002) The single transmembrane domains of ErbB receptors self-associate in cell membranes. J Biol Chem 277(7):4704–4712

14. Laage R, Langosch D (1997) Dimerization of the synaptic vesicle protein synaptobrevin (vesicle-associated membrane protein) II depends on specific residues within the transmembrane segment. Eur J Biochem 249(2): 540–546

15. Huber O, Kemler R, Langosch D (1999) Mutations affecting transmembrane segment interactions impair adhesiveness of E-cadherin. J Cell Sci 112(Pt 23):4415–4423

16. Li R et al (2004) Dimerization of the transmembrane domain of Integrin alphaIIb subunit in cell membranes. J Biol Chem 279(25): 26666–26673

17. McClain MS et al (2003) Essential role of a GXXXG motif for membrane channel formation by Helicobacter pylori vacuolating toxin. J Biol Chem 278(14):12101–12108

18. Bowen ME, Engelman DM, Brunger AT (2002) Mutational analysis of synaptobrevin transmembrane domain oligomerization. Biochemistry 41(52):15861–15866

19. Russ WP, Engelman DM (2000) The GxxxG motif: a framework for transmembrane helix-helix association. J Mol Biol 296(3):911–919

20. Finger C, Escher C, Schneider D (2009) The single transmembrane domains of human receptor tyrosine kinases encode self-interactions. Sci Signal 2(89):ra56

21. Dawson JP, Weinger JS, Engelman DM (2002) Motifs of serine and threonine can drive association of transmembrane helices. J Mol Biol 316(3):799–805

22. Lindner E et al (2007) An extended ToxR POSSYCCAT system for positive and negative selection of self-interacting transmembrane domains. J Microbiol Methods 69(2):298–305

23. Herrmann JR et al (2010) Ionic interactions promote transmembrane helix-helix association depending on sequence context. J Mol Biol 396(2):452–461

24. Lindner E, Langosch D (2006) A ToxR-based dominant-negative system to investigate heterotypic transmembrane domain interactions. Proteins 65(4):803–807

25. Gerber D, Shai Y (2001) In vivo detection of hetero-association of glycophorin-A and its mutants within the membrane. J Biol Chem 276(33):31229–31232

26. Markham BE, Little JW, Mount DW (1981) Nucleotide sequence of the lexA gene of Escherichia coli K-12. Nucleic Acids Res 9(16):4149–4161

27. Horii T et al (1981) Regulation of SOS functions: purification of E. coli LexA protein and determination of its specific site cleaved by the RecA protein. Cell 27(3 Pt 2):515–522

28. Schnarr M et al (1991) DNA binding properties of the LexA repressor. Biochimie 73(4): 423–431

29. Porte D et al (1995) Fos leucine zipper variants with increased association capacity. J Biol Chem 270(39):22721–22730

30. Dmitrova M et al (1998) A new LexA-based genetic system for monitoring and analyzing protein heterodimerization in Escherichia coli. Mol Gen Genet 257(2):205–212

31. Schneider D, Engelman DM (2004) Involvement of transmembrane domain interactions in signal transduction by alpha/beta integrins. J Biol Chem 279(11):9840–9846

32. Schneider D, Engelman DM (2004) Motifs of two small residues can assist but are not sufficient to mediate transmembrane helix interactions. J Mol Biol 343(4):799–804

33. Escher C, Cymer F, Schneider D (2009) Two GxxxG-like motifs facilitate promiscuous interactions of the human ErbB transmembrane domains. J Mol Biol 389(1):10–16

34. King G, Dixon AM (2010) Evidence for role of transmembrane helix-helix interactions in the assembly of the Class II major histocompatibility complex. Mol Biosyst 6(9): 1650–1661

35. Prodohl A et al (2005) Defining the structural basis for assembly of a transmembrane cytochrome. J Mol Biol 350(4):744–756

36. Cymer F, Schneider D (2009) A single glutamate residue controls the oligomerisation, function and stability of the aquaglyceroporin GlpF. Biochemistry 49:279–286

37. Prodöhl A et al (2007) A mutational study of transmembrane helix-helix interactions. Biochimie 89(11):1433–1437

38. Finger C et al (2006) The stability of transmembrane helix interactions measured in a biological membrane. J Mol Biol 358(5):1221–1228

39. Karimova G, Ullmann A, Ladant D (2001) Protein-protein interaction between Bacillus stearothermophilus tyrosyl-tRNA synthetase subdomains revealed by a bacterial two-hybrid system. J Mol Microbiol Biotechnol 3(1):73–82

40. Gropp M et al (2001) Regulation of Escherichia coli RelA requires oligomerization of the C-terminal domain. J Bacteriol 183(2): 570–579

41. Lee H et al (2001) SeqA protein aggregation is necessary for SeqA function. J Biol Chem 276(37):34600–34606

42. Lehnik-Habrink M et al (2011) RNase Y in Bacillus subtilis: a natively disordered protein that is the functional equivalent of RNase E from Escherichia coli. J Bacteriol 193(19): 5431–5441

43. Luo ZQ, Isberg RR (2004) Multiple substrates of the Legionella pneumophila Dot/Icm system identified by interbacterial protein transfer. Proc Natl Acad Sci USA 101(3): 841–846

44. Duerig A et al (2009) Second messenger-mediated spatiotemporal control of protein degradation regulates bacterial cell cycle progression. Genes Dev 23(1):93–104

45. Ladant D, Ullmann A (1999) Bordatella pertussis adenylate cyclase: a toxin with multiple talents. Trends Microbiol 7(4):172–176

46. Karimova G et al (1998) A bacterial two-hybrid system based on a reconstituted signal transduction pathway. Proc Natl Acad Sci USA 95(10):5752–5756

47. Ullmann A, Danchin A (1983) Role of cyclic-Amp in bacteria. Adv Cyclic Nucleotide Res 15:1–53

48. Karimova G, Ullmann A, Ladant D (2000) A bacterial two-hybrid system that exploits a cAMP signaling cascade in Escherichia coli. Methods Enzymol 328:59–73

49. Karimova G, Dautin N, Ladant D (2005) Interaction network among Escherichia coli membrane proteins involved in cell division as revealed by bacterial two-hybrid analysis. J Bacteriol 187(7):2233–2243

50. Karimova G, Robichon C, Ladant D (2009) Characterization of YmgF, a 72-residue inner membrane protein that associates with the Escherichia coli cell division machinery. J Bacteriol 191(1):333–346

51. Engelman DM et al (2003) Membrane protein folding: beyond the two stage model. FEBS Lett 555(1):122–125

52. Cymer F, Veerappan A, Schneider D (1818) Transmembrane helix-helix interactions are modulated by the sequence context and by lipid bilayer properties. Biochim Biophys Acta 4:963–973

53. Anbazhagan V, Schneider D (2010) The membrane environment modulates self-association of the human GpA TM domain—implications for membrane protein folding and transmembrane signaling. Biochim Biophys Acta 1798(10):1899–1907

54. Anbazhagan V, Cymer F, Schneider D (2010) Unfolding a transmembrane helix dimer: a FRET study in mixed micelles. Arch Biochem Biophys 495(2):159–164

55. Sambrook J, Russell DW (2001) Molecular cloning: a laboratory manual, 3rd edn. Cold Spring Harbor Laboratory Press, New York

56. Kolmar H et al (1994) Dimerization of Bence Jones proteins: linking the rate of transcription from an Escherichia coli promoter to the association constant of REIV. Biol Chem Hoppe Seyler 375(1):61–70

57. Duplay P, Szmelcman S (1987) Silent and functional changes in the periplasmic maltose-binding protein of Escherichia coli K12. II. Chemotaxis towards maltose. J Mol Biol 194(4):675–678

58. Treptow NA, Shuman HA (1985) Genetic evidence for substrate and periplasmic-binding-protein recognition by the MalF and MalG proteins, cytoplasmic membrane components

of the Escherichia coli maltose transport system. J Bacteriol 163(2):654–660

59. Kolmar H et al (1995) Immunoglobulin mutant library genetically screened for folding stability exploiting bacterial signal transduction. J Mol Biol 251(4):471–476

60. Herrmann JR et al (2009) Complex patterns of histidine, hydroxylated amino acids and the GxxxG motif mediate high-affinity transmembrane domain interactions. J Mol Biol 385(3): 912–923

61. Lemmon MA et al (1992) Sequence specificity in the dimerization of transmembrane alpha-helices. Biochemistry 31(51):12719–12725

62. Sulistijo ES, Jaszewski TM, MacKenzie KR (2003) Sequence-specific dimerization of the transmembrane domain of the "BH3-only" protein BNIP3 in membranes and detergent. J Biol Chem 278(51):51950–51956

63. Duplay P et al (1987) Silent and functional changes in the periplasmic maltose-binding protein of Escherichia coli K12. I. Transport of maltose. J Mol Biol 194(4):663–673

64. Bormann BJ, Knowles WJ, Marchesi VT (1989) Synthetic peptides mimic the assembly of transmembrane glycoproteins. J Biol Chem 264(7):4033–4037

65. Schneider D et al (2007) From interactions of single transmembrane helices to folding of alpha-helical membrane proteins: analyzing transmembrane helix-helix interactions in bacteria. Curr Protein Pept Sci 8(1):45–61

Part II

Interactions with the Lipid Bilayer

Chapter 5

Analyzing the Effects of Hydrophobic Mismatch on Transmembrane α-Helices Using Tryptophan Fluorescence Spectroscopy

Gregory A. Caputo

Abstract

Hydrophobic matching between transmembrane protein segments and the lipid bilayer in which they are embedded is a significant factor in the behavior and orientation of such transmembrane segments. The condition of hydrophobic mismatch occurs when the hydrophobic thickness of a lipid bilayer is significantly different than the length of the membrane spanning segment of a protein, resulting in a mismatch. This mismatch can result in altered function of proteins as well as nonnative structural arrangements including effects on transmembrane α-helix tilt angles, oligomerization state, and/or the formation of non-transmembrane topographies. Here, a fluorescence-based protocol is described for testing model transmembrane α-helices and their sensitivity to hydrophobic mismatch by measuring the propensity of these helices to form non-transmembrane structures. Overall, good hydrophobic matching between the bilayer and transmembrane segments is an important factor that must be considered when designing membrane proteins or peptides.

Key words Fluorescence spectroscopy, Fluorescence quenching, Hydrophobic mismatch, Transmembrane helices, Peptide design

1 Introduction

Membrane proteins have been notoriously difficult to work with, especially when performing sensitive biophysical and high resolution structural studies. However, advances in spectroscopic and crystallographic techniques have resulted in a tremendous increase in the number of high resolution membrane protein structures available as well as increased understanding of how these membrane proteins behave in the lipid environment to perform their intended functions. Although still lagging behind soluble proteins, the increase in fundamental understanding of membrane protein biophysics and mechanisms has enabled the knowledge-based design of membrane proteins to tackle significant design problems including the de novo design of functional transmembrane (TM) proteins and peptides [1–5].

Giovanna Ghirlanda and Alessandro Senes (eds.), *Membrane Proteins: Folding, Association, and Design*,
Methods in Molecular Biology, vol. 1063, DOI 10.1007/978-1-62703-583-5_5, © Springer Science+Business Media, LLC 2013

There are numerous factors that must be considered in the design of transmembrane (TM) protein structures [2–5]. These include amino acid composition, functional sequence motifs such as oligomerization motifs or flanking residues, propensity to form the desired secondary structure, and the length of the hydrophobic sequence. Each of these design components relates to one another, resulting in a very fine balance required for functional designs. Numerous reports have demonstrated proteins which show a direct relationship between hydrophobic mismatch and function [6–10]. The length of the protein segment required to span the bilayer generally falls within a narrow window of lengths which is dictated by the thickness of the lipid bilayer [11–13]. However, changes in oligomerization state and tilt angle in the bilayer can impact both function and sensitivity to mismatch [14, 15]. Hence, careful design of hydrophobic length is critical for TM structures in order to avoid hydrophobic mismatch.

When discussing transmembrane proteins, hydrophobic mismatch can be generally defined as a condition in which the hydrophobic thickness of the lipid bilayer does not match the hydrophobic length of the protein segment intended to span the bilayer [14, 16]. This can occur when the hydrophobic length of the TM segment is greater than the hydrophobic thickness of the bilayer or, conversely, when the hydrophobic length of the TM segment is shorter than the thickness of the bilayer. These conditions, referred to as positive and negative mismatch, respectively, are schematized in Fig. 1.

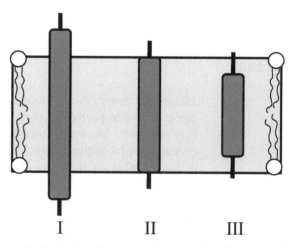

I   II   III

Fig. 1 Schematic illustration of hydrophobic mismatch with transmembrane helices. The *rectangle* (*blue*) represents the hydrophobic portion of the α-helix that favorably partitions in the nonpolar core of the lipid bilayer (*gray*). The three typical conditions of mismatch are (I) negative mismatch, (II) no mismatch (optimal matching), and (III) positive mismatch. Negative mismatch results in hydrophobic amino acids from the helix being exposed to water while positive mismatch results from hydrophobic helix length insufficient to span the bilayer

The effects of hydrophobic mismatch on naturally occurring membrane proteins have been well documented. The effects range from alterations in channel gating in the case of the large mechanosensitive channel of *E. coli*, transporter activity in the Leu-H+ cotransporter of *L. lactis*, enzymatic activity of cytochrome C oxidase, antibiotic binding to the F1-Fo ATPase, as well as sorting of proteins in the secretory pathway [8, 9, 13, 17–19]. It is important to note that these mismatch-sensitive proteins are found in numerous different organisms as well as located in both cellular membranes and internal organellar membranes. This implies that the effect of mismatch on these proteins is not a specific interaction but a more general, physical principle that changes the inherent properties of the protein embedded in the bilayer.

A significant amount of work on the fundamental principles of hydrophobic mismatch has been carried out using model hydrophobic peptides which form TM α-helices [16, 20–26]. This is primarily due to the added level of difficulty in expression, purification, and functional reconstitution of membrane proteins in model membrane vesicles compared to the peptide counterparts. Additionally, advances in solid phase peptide synthesis have allowed for relative ease in synthesis of hydrophobic peptides on the length scale of most normal TM proteins. These studies have investigated numerous specific peptide sequences and often relate aspects of amino acid composition to mismatch responses [16, 21, 22].

There are several classes of helical TM peptides that have been well studied regarding the responses to hydrophobic mismatch. One of these is the WALP class of sequences which derive their name from the Trp (W) terminal anchors and the alternating Ala (A) and Leu (L) amino acids that make up the hydrophobic core of the TM helix [20, 27–29]. Related peptide sequences use the alternating Ala-Leu TM domain but flank the peptide with cationic lysines to serve as anchoring residues (KALP) [28, 30]. Several groups have shown that these WALP and KALP sequences adapt to hydrophobic mismatch through a variety of mechanisms including bilayer modulation, tilting, and backbone flexibility [20, 27–29, 31, 32]. Another well studied class of model TM helices is based on a poly-leucine TM domain [16, 23, 33–40]. These sequences provide a more stable TM helix due to the increased hydrophobicity which provides greater resistance to adopting non-bilayer conformations compared with the Ala-Leu hydrophobic domains. The increased hydrophobicity of poly-Leu sequences also provides a more stable TM helix framework for the investigation of polar and ionizable residues in the TM domain [16, 38]. This is an important consideration in that mismatch responses are known to be impacted by primary sequence [16, 24, 38, 41].

The most common approach to study hydrophobic mismatch centers on the creation of model membrane vesicles composed of lipids with varying length acyl chains. The length of the acyl chains of the lipids is the key determinant of the bilayer hydrophobic thickness where extending the acyl chain linearly increases the thickness of the bilayer as a function of the number of additional carbon atoms in the acyl chain [42–45]. In vesicles that are formed with lipids containing short acyl chains, the hydrophobic core of the bilayer is narrow or "thin" while longer acyl chain lipids create wide or "thick" bilayers. This direct relationship between acyl chain length and bilayer thickness allows for simple experimental investigation of hydrophobic mismatch using commercially available lipid species [16, 24, 38, 40]. There are alternative methods to vary bilayer hydrophobic thickness including the incorporation/ removal of cholesterol, incorporation/removal of long chain alcohols that partition into the bilayer, and induction of phase transitions in the bilayer [16, 46]. However, these methods do not allow precise control over bilayer thickness nor do they allow for as great of a range of bilayer thicknesses to be investigated.

There are a variety of methods that can be employed to study the effects of hydrophobic mismatch on TM peptide behavior. Protein-specific functional assays for membrane proteins and the well characterized glycosylation mapping method can yield important information on mismatch, but are inappropriate or incompatible with model TM helices [47–51]. Instead, the bulk of information regarding peptide responses to mismatch has come from a wide variety of spectroscopic techniques as well as physical methods such as calorimetry and X-ray scattering. These investigations can be separated into those which measure the changes in the lipid properties in response and those which interrogate the peptide.

Differential scanning calorimetry (DSC) can readily measure changes in lipid bilayer properties through changes in phase transition temperatures, indicative in changes in lipid packing within the bilayer. These studies have consistently shown that mismatch alters the main chain melting transition, generally causing an upward shift in melting temperature (T_m) for positive mismatch conditions and a downward shift for negative mismatch [23, 33, 52]. Alternatively, X-ray and neutron scattering studies can accurately measure lipid bilayer thicknesses and can be useful in the identification of non-bilayer phase lipids in the mismatch response [27, 43, 44, 53, 54]. While the majority of spectroscopic investigations of TM helices focus on interrogation of the peptide, ^{31}P NMR can be used to interrogate phase transition temperatures as well as report on specific lipid phase behavior including the formation of non-bilayer structures [20, 32, 55, 56].

Spectroscopic investigations of the peptides in mismatch conditions are often designed to directly report the orientation or topology of the peptide with respect to the bilayer. These often focus on distinguishing between helices that adopt TM structures and those that adopt a non-TM or membrane surface topology in response to the mismatch conditions or a change in tilt angle. Fluorescence, electron paramagnetic resonance (EPR), NMR, oriented circular dichroism (OCD), and polarized attenuated total reflectance infrared (pATR-IR) each can report on peptide topography. EPR spectroscopy frequently uses the interaction of spin-labeled groups as reporters of environment and protein–protein interactions. Through measuring the degree of interaction of spin-labeled groups attached to TM helices, the response to mismatch by changing helix tilt angle has been successfully determined [57, 58]. NMR, OCD, and pATR-IR can directly measure the orientation of a helix in a membrane environment and can be used to directly calculate the angle at which a TM helix is inserted in the bilayer, commonly referred to as the tilt angle [56, 59–71]. Through changes in spectral shape, OCD spectra can be interpreted to determine if a given helix is parallel to the plane of the bilayer (lying on the surface of the bilayer) or perpendicular to the bilayer plane (TM orientation) [62–66]. Similarly, pATR-IR directly interrogates the topology of the helix by utilizing differential absorptivities of parallel and perpendicularly polarized incident radiation [67, 69–74]. A major drawback in both OCD and pATR-IR is the absolute requirement for well oriented lipid bilayers (or monolayers), generally achieved by the creation of supported lipid multilayers. These supported bilayers do not perfectly mimic natural bilayers and are often partially dehydrated during the measurements which can impact energetic interactions driven by hydrophobicity. Nonetheless, these measurements are of great benefit in understanding topographic responses to mismatch or other environmental conditions.

Fluorescence spectroscopy can be applied to several aspects of mismatch studies including oligomerization, topography, and lipid behavior. Energy transfer studies are routinely used to measure the association of proteins or peptides in membranes and can be useful in identifying oligomers induced by either positive or negative mismatch [75–78]. Fluorescence quenching methods have been applied to oligomerization of TM peptides and can also easily distinguish between bilayer-embedded fluorophores and those exposed to the aqueous milieu, although this is dependent on appropriate fluorophore/quencher pairs and the location of the fluorophore within the sequence [16, 24, 38–41, 46, 79–83].

However, the most straightforward application of fluorescence spectroscopy to hydrophobic mismatch is topographical studies using an environmentally sensitive probe. These probes will exhibit altered emission properties when in an aqueous environment (the bilayer surface) or in a nonpolar environment (the bilayer core). Through incorporation of the fluorescent probe at the center of the peptide sequence in question, the transition from TM to non-TM topography will result in a dramatic change in local polarity around the fluorophore resulting in a spectral shift [16, 24, 38–41, 46, 79, 80]. The naturally occurring fluorophore Trp undergoes this type of environmentally mediated spectral shift, exhibiting a red-shifted emission spectrum when at the membrane surface ($\lambda_{max} \approx 335$ nm) and a significantly blue-shifted emission when buried at the bilayer center ($\lambda_{max} \approx 320$ nm) [16, 24, 38–41, 46, 79, 80]. This property of Trp allows for the direct interrogation of peptide topography with fluorescence spectroscopy without the need for exogenous labels. Appropriate incorporation of a Trp into a designed peptide sequence will ensure the Trp is located at or near the bilayer center in a TM orientation, ensuring a large shift in λ_{max} if the peptide does not form a stable TM helix. It should be noted that studies on the positional preferences of amino acids in membranes indicate Trp has a strong preference for interfacial locations but this preference is not sufficient to prevent an otherwise hydrophobic helix from adopting a stable TM orientation, even when the resulting helix positions the Trp near the bilayer center [16, 84–87].

As described, modulating the length of the lipid acyl chains in a bilayer inherently alters the hydrophobic thickness of that bilayer, potentially creating positive or negative hydrophobic mismatch conditions with a given TM helix [42–45]. These mismatch conditions could be negative mismatch if the acyl chains are short, resulting in a thin bilayer, or a positive mismatch condition when acyl chains are long, resulting in a thicker bilayer. Positive and negative mismatch can induce multiple different conformational responses in TM helices (Fig. 2) and may necessitate additional investigations to determine exact orientations or conformations [14, 27, 88].

As a TM helix in question changes orientation or conformation, in most cases the Trp in the sequence will undergo a change in environment such that a spectral shift will be observed. Under the conditions of positive mismatch (hydrophobic helix > hydrophobic bilayer thickness), TM helices have been shown to tilt or form oligomeric structures as well as induce changes in the lipid bilayer structure (Fig. 2a) [16, 21, 25, 27, 57, 59–61, 89–91]. Helix tilting to compensate for mismatch maximizes the burial of as much hydrophobic bulk in the bilayer core [57, 59, 60, 89, 90, 92]. All of these structures in thin bilayers result in a Trp buried in the acyl chain region of the bilayer but relatively close to the bilayer surface due to the shorter length of the acyl chains [16, 24, 38–41, 79]. When the helix length is near or equal to the thickness of the bilayer, this is deemed to be optimal matching (or lack of mismatch) resulting in a stable TM orientation. Adopting this TM orientation, the Trp should be buried in the nonpolar core of the bilayer and result in a blue-shifted emission spectrum [16, 24, 38–41, 79]. In the case of negative mismatch conditions (hydrophobic helix < hydrophobic bilayer thickness), helices have also been shown to oligomerize, alter their structure, induce changes in the bilayer, and reorient with respect to the bilayer (Fig. 2b) [11, 14, 16, 24, 27, 38–41, 46, 91–94]. If the peptide cannot compensate for the energetic penalty of mismatch, it may adopt a non-bilayer topography bringing the Trp into the more polar, aqueous environment of the bilayer while still burying some of the hydrophobic side

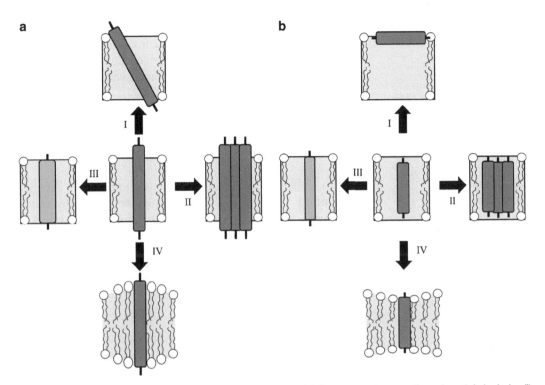

Fig. 2 Responses of TM α-helices to mismatch conditions. (**a**) Responses to *negative* mismatch include: (I) tilting to bury more hydrophobic amino acids in the bilayer, (II) oligomerization to bury hydrophobic amino acids in protein–protein contacts, (III) deformation of the lipid bilayer to accommodate the hydrophobic helix, or (IV) alteration of the secondary structure of the protein/peptide to facilitate matching (shown as a *green* hydrophobic segment). (**b**) Responses to *positive* mismatch include: (I) adopting a non-transmembrane topography, (II) oligomerization to bury hydrophobic amino acids in protein–protein contacts, (III) deformation of the lipid bilayer to accommodate the hydrophobic helix, or (IV) alteration of the secondary structure of the protein/peptide to facilitate matching (shown as a *green* hydrophobic segment)

chains. This shallower location of the Trp results in a more red-shifted fluorescence emission spectrum [16, 24, 38–41, 79, 95].

Below, the standard protocol used to screen single-pass TM helices for sensitivity and response to hydrophobic mismatch is outlined. This procedure takes advantage of the ease of creating bilayers of varying hydrophobic thickness using model membrane vesicles and the environmentally sensitive fluorophore tryptophan.

2 Materials

2.1 Lipid Stock Solutions

1. Lipids (Avanti Polar Lipids, Alabaster, AL) as lyophilized powders, or dissolved in chloroform. Lipids should be stored in glass vials at either –20 °C (working solutions) or –70 to –80 °C (stock solutions).

2. Aluminum foil. To shield samples from light.

3. A balance accurate to 0.01 mg. A Cahn C-33 or a Perkin-Elmer AD-6 Microbalance was used for our studies. A standard milligram scale balance can be used, but will require higher amounts of lipid to obtain samples within the mass tolerances of the scale (*see* **Note 1**).

4. Nitrogen, argon, or other inert gas source with regulator capable of producing a gentle stream (~2 psi).

5. Vacuum desiccator.

2.2 Sample Preparation

1. Pipettes which are organic solvent-compatible. We use Drummond positive-displacement pipettes with disposable glass bores. Changing bores between samples is essential to avoid cross-contamination of different lipid species.

2. Borosilicate glass test tubes.

3. Nitrogen or other inert gas source with regulator capable of producing a gentle stream (~2 psi).

4. Vortexing mixer.

5. A dark place to store samples before measurement.

2.3 Fluorescence Measurements

1. Monochromator-based fluorometer.

2. Quartz cuvettes (*see* **Note 2**).

3. Standard pipettes.

3   Methods

3.1 Experimental Outline

The experiments described use multiple sets of samples to measure differences in fluorescence emission properties as a function of lipid bilayer thickness. Depending on the extent of the mismatch conditions, the thermodynamic penalty for mismatch could overwhelm the favorable energy of helix insertion into the membrane. The energetics of the interplay between mismatch and TM helix formation is inherently linked to the amino acid sequence, both length and identity, that is interacting with the membrane. The method outlined is intended as an initial screen of helix responses to positive and negative mismatch as reported by changes in the fluorescence emission spectrum of the environmentally sensitive amino acid Trp. The procedure results in a plot of Trp emission λ_{max} as a function of bilayer acyl chain length which is indicative of how the helix responds to a range of mismatch conditions [16, 38–40].

Six independent sets of samples are prepared in at least triplicate, each containing the hydrophobic peptide in question and one of the lipid species being investigated. Our studies have used lipids with symmetric acyl chains, each chain bearing one desaturation, to ensure uniform thickness and bilayer fluidity at room temperature [16, 38–40]. The acyl chain lengths range from 14 to 24C which is the entire range of lengths that is readily commercially available. This range of thicknesses brackets the most common naturally occurring acyl chain lengths (16 and 18C), allowing for examination of both positive and negative mismatch. However, other lengths are available as saturated chains and mixed-acyl chain lipids if necessary (*see* **Note 3**). Single background samples lacking the peptide are prepared for each lipid species being investigated. Samples are excited at 280 nm and the fluorescence emission spectra are recorded over the range 300–375 nm (*see* **Note 4**). The emission spectra are then analyzed to determine λ_{max} (the wavelength of maximum emission intensity) and these λ_{max} values can be plotted as a function of acyl chain length [16, 38–40].

3.2 Lipid Stock Solutions

Significant caution must be given to preparation of accurate lipid stock solutions to ensure accurate peptide:lipid ratios in the samples. This is of critical importance in that many sequences may show some tendency to oligomerize under mismatch conditions, a situation that could be exacerbated if the peptide:lipid ratio is too low, due to improper determination of lipid concentrations. Additionally, sequences may be in equilibrium between TM and non-TM conformations under specific mismatch conditions and that equilibrium position could be shifted by changes in peptide:lipid ratio. Lipid stock concentrations are generally performed upon creation of a new lipid stock (*see* **Note 5**).

1. Lipids are purchased as lyophilized solids or dissolved in chloroform. Solid samples can be subsequently dissolved in chloroform. Lipid stock solutions are stored in glass vials at –20 °C while the lyophilized solids are generally stored at –80 °C. Stock vials are additionally sealed by wrapping the lid of the bottle or vial with Teflon tape and/or parafilm to minimize contamination.

2. Aluminum foil is cut into small squares, numbered on the exterior surface, and then formed into a cup-shape using forceps. Alternatively, the caps of 1.5 mL microcentrifuge tubes can be cut off and directly used as cups. This method is more straightforward but the increased weight of these caps may exceed the measurable range of some microbalances (see below).

3. The aluminum foil cups or centrifuge tube caps are weighed on a microbalance to the closest microgram. This should be done using gloved hands and forceps. The individual weights are then recorded (*see* **Note 6**).

4. Next 25–50 μL of lipid solution is gently pipetted into the cup. For best accuracy, 3–5 replicates are created.

5. The sample is dried by evaporation of the solvent under a gentle stream of nitrogen (<2 psi) for 10 min to create lipid films. At this step, significant care should be taken to avoid splashing or splattering of the lipid solution out of the cups which will

result in inaccurate quantification of lipid concentrations. Complete drying of the lipids into a film absent of any solvent is essential for accurate weight determination.

6. The lipid films are placed in a desiccator jar under high vacuum for 45–90 min to ensure complete removal of any residual solvents. Typically we place all of the cups in a small beaker to aid in addition and removal from the desiccator jar.

7. The cups are then reweighed and the difference in weight is used to calculate the concentration of the solution (*see* **Note 7**).

3.3 Peptide Purification and Stock Preparation

Peptide sequences corresponding to a TM helix of interest are generally synthesized by solid phase methods. This ideally produces a fluffy white crude peptide sample. Solubilization and purification conditions will vary depending on the specific sequences being studied, but a general outline is provided below [80]:

1. Crude peptide is dissolved in a mixture of water and nonpolar solvent (acetonitrile, isopropanol, or an acetonitrile/isopropanol mixture) supplemented with 0.1–0.5 % TFA. The peptide should be dissolved in a solution composed of as little organic solvent as possible to maintain solubility of the mixture (*see* **Note 8**).

2. The crude peptide solution is filtered using a 0.22 μm syringe filter.

3. This mixture is injected onto an HPLC column equilibrated in the same solvent mixture in which the peptide is dissolved. A reversed phase column with a stationary phase of C3, C4, C8, or C18 was used for purification of these sequences.

4. The hydrophobic peptide is eluted using a linear gradient of water and nonpolar solvent, typically increasing the nonpolar solvent at 1 % per minute of run time.

5. Analyte detection is performed using an absorbance detector set to 220 nm (peptide bond absorbance), 280 nm (Trp absorbance), or both (*see* **Note 9**).

6. Fractions potentially containing the sequence of interest (i.e., corresponding to peaks on the absorbance detector) are analyzed by mass spectrometry to confirm peptide identity.

7. Fractions containing the peptide of interest can be frozen and lyophilized or dried under a gentle stream (~2 psi) of N_2.

8. Purified peptide samples are dissolved in a 1:1 mixture of H_2O:ethanol or H_2O:isopropanol.

9. Concentration of the peptide stock solution is then determined by absorbance spectroscopy using the Trp absorbance at 280 nm ($\varepsilon = 5{,}560$ cm^{-1} M^{-1}).

10. Peptide stock solutions are stored at –20 °C.

3.4 Sample Preparation

Lipid vesicles containing the peptide of interest (as well as background controls) are created using the ethanol dilution method. Briefly, the dried lipid–peptide film is dissolved in a small volume of absolute ethanol and buffer is added rapidly while vortexing. This quick buffer addition to a solution of lipids in ethanol promotes the formation of small unilamellar vesicles (SUVs) [16, 38, 79, 96, 97]. Alternatively, the samples can be prepared as multilamellar vesicles (MLVs) or extruded large unilamellar vesicles (LUVs); however, these vesicle types often result in increased scatter in the measurement and, in the case of extruded LUVs, are much more time-consuming to prepare. The method of ethanol dilution for vesicle preparation is outlined below:

1. An appropriate amount of each lipid stock solution is added to glass test tubes using the organic-compatible pipette. Samples are prepared at least in triplicate. A single background sample for each lipid thickness is also prepared with the omission of the peptide.

2. As the peptide sequences in question are generally strongly hydrophobic, the peptide of interest is next added to each of the samples (but not to the tube designated as the background) from a stock solution in ethanol/water or isopropanol/water. This results in a lipid:peptide mixture in a small volume of primarily organic solvent.

3. Solvents are then evaporated using a gentle stream (~2 psi) of pure nitrogen or other inert gas for approximately 10 min or until the samples are visibly dry. Dried samples will generally appear as a translucent spot at the bottom of the tube. Caution should be taken to avoid splashing of the lipid:peptide mixture onto the walls of the tube due to gas flow.

4. The tubes containing the lipid:peptide films are placed in a vacuum desiccator and held under vacuum for 1–2 h. This step is to ensure complete removal of all traces of organic solvents from the peptide and lipid stocks.

5. Using a positive-displacement pipette, 10 μL of ethanol is added to the thoroughly dried lipid:peptide film. This is immediately swirled by hand or with *gentle* vortexing to ensure complete dissolution of the film without splashing. Complete dissolution of the lipid:peptide film is necessary for the formation of SUVs with the desired lipid:peptide ratio.

6. Upon dissolution of the lipid:peptide film, 790 μL of sample buffer is added to the ethanolic solution of lipid and peptide while vigorously vortexing. Vortexing is continued for 10–15 s after addition of the buffer. Sample volumes can be modified to suit the cuvettes available.

7. Samples are allowed to equilibrate for approximately 30 min. Ideally the incubation should occur where the samples are protected from light. This incubation is generally at room temperature but can be modified depending on the experiment.

8. Prior to measuring fluorescence, samples are vortexed and transferred to quartz cuvettes.

3.5 Fluorescence Measurements

1. Samples are placed into fluorescence cuvettes (polished on all four sides). We use semi-micro quartz fluorescence cuvettes (10 mm excitation path length, 4 mm emission path length) to reduce the required volume and consumption of peptide. The cuvette is inserted into the holder in the sample chamber of the fluorometer.

2. The excitation and emission slits are set at appropriate values (*see* **Note 10**).

3. Emission spectra of all samples and backgrounds are recorded in order to determine the emission λ_{max}. The excitation monochromator is set to 280 nm and fluorescence emission is generally collected over the range 300–400 nm. Spectra should be recorded at 1 nm resolution or higher. The range of wavelengths scanned can be extended if the peptide in question exhibits strongly red-shifted Trp fluorescence (*see* **Note 11**).

3.6 Backgrounds and Controls

As noted, background samples containing only lipid (one for each acyl chain length being tested) should be prepared [16, 24, 38–40, 79]. The apparent emission intensity in these samples is subtracted from the intensity of the samples containing the Trp-containing peptide molecule in order to accurately determine the emission spectrum of the peptide in question. It can be expected that samples containing lipid vesicles will scatter some small amount of the excitation light and appear in the lower end of the emission spectrum. Additionally, when emission spectra are being recorded at an excitation wavelength of 280 nm, a large, sharp peak at ~310 nm arising from Raman scattering from water should be expected.

There are several control experiments that should be performed to ensure proper interpretation of data. Ideally, peptides containing Trp at known positions are used as environmental control samples, yielding model emission spectra for a Trp buried in the bilayer and one for a Trp at the surface of the bilayer [40, 46]. For this purpose, we routinely use Lys-flanked poly-Leu peptides which contain a single Trp, either in the middle of the sequence or near one of the termini, resulting in a deeply buried Trp or a shallowly located Trp. This class of peptide has been extensively studied by spectroscopic methods and has been shown to adopt stable TM α-helical structures [16, 38–40, 46].

Another essential control is confirming that the sequence in question has reached equilibrium with respect to structure, topography, and oligomeric state over the course of the sample preparation. These controls can be performed by simply measuring the emission spectra from a given sample over time. In our experience, 30 min of equilibration time prior to measurement is more than sufficient for poly-Leu sequences; however, longer time scales should be investigated during the initial investigation of a peptide sequence [16, 38–40, 46].

A final set of control experiments may be necessary to ensure directionality of helix insertion in the membrane. In the case of model poly-Leu peptides, the sequences are generally symmetric (or nearly so) eliminating directional impacts. However, extension of the method to naturally derived sequences will inherently require the investigation of asymmetric peptide sequences. The vesicle preparation methods described do not insure unidirectional insertion of the peptide in the bilayer, potentially resulting in a mixed population of structures. Interactions between antiparallel inserted helices may occur in the model vesicle system that would not ordinarily occur in nature. Control experiments for directionality of insertion generally employ chemical reactivity of exogenously added aqueous species with flanking residues. Differences in reactivity profiles may indicate relative populations of heterogeneously oriented helices.

3.7 Data Analysis

All spectra from samples are corrected for background fluorescence from the lipid by simple subtraction. The determination of λ_{max} can be performed visually by inspection of emission intensity as a function of wavelength. In the case of unusually broad or erratic emission spectra, smoothing spectra mathematically can be performed. Additionally, unpublished observations indicate that using spectral barycenter can be equally effective in reporting on shifts in Trp emission properties resulting from changes in bilayer location (*see* **Note 12**).

Upon determination of λ_{max} for each individual sample, the λ_{max} values corresponding to each bilayer thickness are averaged and plotted as a function of bilayer thickness (usually reported as the number of carbons in the acyl chain) [16, 24, 38–41]. These curves have been reported in three general profiles: (I) a U-shape, (II) a flat line, and (III) a shifted U-shape, as shown in Fig. 3 [16, 24, 38–41, 98, 99]. Each of these representative profiles corresponds to different topographical responses to mismatch. In all cases, a blue-shifted λ_{max} is representative of a stable TM helix structure while red shifts are caused in response to positive or negative mismatch. The standard profile exhibited by TM helices is that of the U-shaped curve (I). This shape arises from the Trp being somewhat closer to the membrane surface in short acyl chain bilayers

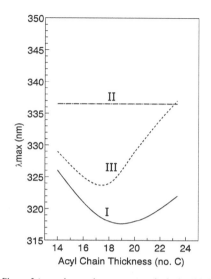

Fig. 3 Typical profiles of λ_{max} dependence on acyl chain thickness. The curves represent a standard, stable TM helix (I), a sequence which does not adopt a TM orientation under any conditions (II), and a shifted helix which is extremely sensitive to mismatch (III). The λ_{max} values reported are representative of typical values previously reported. However, the overall shape of the curves is the distinctive characteristic that informs on helix responses to mismatch. Modified from [38]

(since they are inherently thinner), followed by a blue shift in λ_{max} to the curve's minimum (interpreted as optimal hydrophobic matching as the Trp is buried in the core of the bilayer), followed by a red shift in λ_{max} in long acyl chain bilayers (as mismatch begins to drive helix reorientation to the bilayer surface). Note that the specific bilayer thickness at which the minimum max value occurs is dependent on the length of the hydrophobic helix formed [16, 38, 40, 99]. The second profile, a flat line (II), corresponds to a TM helix that is completely insensitive to the mismatch conditions tested. This could arise from a helix that is stably inserted in the bilayer in all cases or one that is stably associated with the surface of the membrane [16, 38, 95, 99]. Since the peptide does not form a TM structure, the thickness of the bilayer does not impact the stability of the surface-associated conformation. The third observed profile, a shifted U curve (III), is more uncommon but can be attributed to a marginally stable TM helix in thin bilayers but is unable to compensate for significant hydrophobic mismatch and subsequently reorients to the bilayer surface upon exposure to mismatch conditions [16, 24, 38, 41, 95, 99]. This has been observed in the case of helices which contain ionizable residues in the core of the hydrophobic domain resulting in a "shifted" helical segment. Not surprisingly, this shift creates a less stable TM structure which is very sensitive to hydrophobic mismatch, hence the steep response in λ_{max} to changing bilayer thickness. It should be noted that the absolute values of λ_{max} observed in these experiments can vary based on the depth of the Trp when in the TM structure, the orientation of the Trp when in a surface conformation, and Trp interactions with neighboring amino acids [16, 24, 38, 41, 95, 99, 100]. Due to these environmental influences, determination of mismatch responses should be more influenced by the overall shape of the curves obtained rather than on absolute values of λ_{max}.

3.8 Modifications and Extensions of the Procedure

The procedure outlined above is designed to be an initial screen of TM helix response to different degrees of hydrophobic mismatch. As noted above, there are a variety of possible responses to hydrophobic mismatch, many of which can be combinatorial or linked to one another. In these cases, the experimental results for a given peptide sequence may warrant additional experiments to fully understand the mismatch response.

A common extension to test for mismatch sensitivity is in situ bilayer thickness modification through the addition of decane (to increase the bilayer thickness) or removal of cholesterol by β-cyclodextrin (to decrease bilayer thickness) [42, 46, 101–105]. While these methods do not alter the overall thickness significantly, they can be used to tease out small differences in mismatch tolerance or for those cases where the mismatch curve minimum appears to be at an acyl chain length where appropriate lipids may not be available.

In addition to in situ modifications, confirmations of peptide topography under different mismatch conditions are required. It has been shown that λ_{max} can be sensitive to environment, orientation of the peptide, and interactions with neighboring amino acids. As such, the thickness curves alone are insufficient to fully characterize mismatch dependence. These confirmations can be performed using numerous fluorescence techniques utilizing the Trp as the fluorophore (quenching or fluorescence lifetime), OCD, or pATR-IR spectroscopy [16, 46, 56, 65, 69, 71, 72, 106]. Numerous fluorescence quenching methods have been described in the literature to determine fluorophore depth in the membrane including using aqueous quenchers, bilayer-embedded quenchers, or a combination of both [39, 46, 81, 83, 107]. As the Trp reorients from a nonpolar environment in the bilayer core to the aqueous milieu at the bilayer surface, there is also a measurable, concomitant change in the emission lifetime, τ, which can also confirm a change in topography [108]. OCD and pATR-IR rely on differences in CD or IR absorbance spectra that are dependent on the orientation of the helix in supported lipid bilayer. As mentioned earlier, these methods directly interrogate peptide orientation but have the drawback of requiring solid lipid films and not vesicles in solution.

Overall this method is designed as an initial screen to determine sensitivity of a TM helix to hydrophobic mismatch. The use of Trp fluorescence as a reporter of environment negates the need for exogenous probes that can impact behavior as well as the need for solid phase-supported bilayers. The method allows for rapid screening of helical targets and ease of incorporation with other standard spectroscopic techniques used to characterize the behavior of designed TM α-helices.

4 Notes

1. It should be noted that higher amounts of lipids increase the risk of incomplete drying which impacts proper calculation of concentration.

2. High quality quartz cuvettes are required. The excitation and emission wavelengths are in the UV-region and the cuvettes must be transparent in this wavelength region. Either a single cuvette is used or a matched set of cuvettes, meaning that a solution placed in different cuvettes would have the same fluorescence intensity.

3. Lipids for mismatch studies should be in the liquid disordered phase to ensure helices can move and reorient independent of lipid phase artifacts. Use of very long acyl chain lipids or saturated acyl chains may require altering the temperature at which the experiments are performed to ensure proper phase behavior.

4. When investigating peptide sequences which result in significantly red-shifted emission spectra, the range of collection should be extended to ensure complete collection of the emission properties for analysis. For peptides which simply undergo a TM to surface topographical rearrangement, Trp emission spectra usually do not shift past ~330 nm, and hence do not require spectra collection past 375 nm.

5. Lipids should be stored at either –80 °C (stock solutions) or –20 °C (working solutions). Care should be given to ensuring lipids, in powder or liquid form, remain moisture-free. We recommend purging the storage vessels with N_2 briefly and wrapping the vials in parafilm (or similar) before storing. Additionally, minimizing the time the stock solutions are out of the freezer will also prevent absorption of atmospheric water.

6. After weighing, care should be given to not touch the cups to minimize the transfer of hand oils. Precautions include wearing gloves and handling aluminum cups with forceps.

7. Three to five separate samples for dry weight measurements are prepared and weighted for each solution and an average is calculated.

8. Typically a trial dissolution of crude peptide is necessary to determine the minimum amount of organic solvent required. For very hydrophobic sequences that are very poorly soluble in HPLC solvents, dissolution in neat TFA followed by immediate

injection onto the column can be performed. However, this runs the risk of aggregation on the column and difficult elution.

9. High quality HPLC grade solvents are required to minimize the baseline absorbance. This is especially important when monitoring at 220 nm.

10. The specific slit settings of the fluorometer are dependent on the instrument being used. You may have to adjust the slit widths to get a usable signal.

11. Recording entire emission spectra is only strictly required if spectral shift ($\Delta\lambda_{max}$) analysis is being investigated. However, we have often found it informative to have spectral information to ensure comparable behavior in samples from day to day as λ_{max} is more consistent than raw intensity values.

12. Alternatively, spectral barycenter can be calculated for each sample and the total change in barycenter (ΔBc) can also be used to measure heterogeneity (Caputo, unpublished observations).

References

1. Fagerberg L, Jonasson K, von Heijne G, Uhlen M, Berglund L (2010) Prediction of the human membrane proteome. Proteomics 10(6): 1141–1149. doi:10.1002/pmic.200900258

2. Grigoryan G, Moore DT, DeGrado WF (2011) Transmembrane communication: general principles and lessons from the structure and function of the M2 proton channel, K(+) channels, and integrin receptors. Annu Rev Biochem 80:211–237. doi:10.1146/annurev-biochem-091008-152423

3. Perez-Aguilar JM, Saven JG (2012) Computational design of membrane proteins. Structure 20(1):5–14. doi:10.1016/j.str.2011.12.003

4. Senes A, Engel DE, DeGrado WF (2004) Folding of helical membrane proteins: the role of polar, GxxxG-like and proline motifs. Curr Opin Struct Biol 14(4):465–479. doi:10.1016/j.sbi.2004.07.007

5. Slivka PF, Wong J, Caputo GA, Yin H (2008) Peptide probes for protein transmembrane domains. ACS Chem Biol 3(7):402–411. doi:10.1021/cb800049w

6. Caffrey M, Feigenson GW (1981) Fluorescence quenching in model membranes. 3. Relationship between calcium adenosinetriphosphatase enzyme activity and the affinity of the protein for phosphatidylcholines with different acyl chain characteristics. Biochemistry 20(7):1949–1961

7. Deshmukh SS, Akhavein H, Williams JC, Allen JP, Kalman L (2011) Light-induced conformational changes in photosynthetic reaction centers: impact of detergents and lipids on the electronic structure of the primary electron donor. Biochemistry 50(23): 5249–5262. doi:10.1021/bi200595z

8. In't Veld G, Driessen AJ, Op den Kamp JA, Konings WN (1991) Hydrophobic membrane thickness and lipid-protein interactions of the leucine transport system of Lactococcus lactis. Biochim Biophys Acta 1065(2):203–212

9. Montecucco C, Smith GA, Dabbeni-sala F, Johannsson A, Galante YM, Bisson R (1982) Bilayer thickness and enzymatic activity in the mitochondrial cytochrome c oxidase and ATPase complex. FEBS Lett 144(1):145–148

10. Pilot JD, East JM, Lee AG (2001) Effects of bilayer thickness on the activity of diacylglycerol kinase of Escherichia coli. Biochemistry 40(28):8188–8195

11. Andersen OS, Koeppe RE II (2007) Bilayer thickness and membrane protein function: an energetic perspective. Annu Rev Biophys Biomol Struct 36:107–130. doi:10.1146/annurev.biophys.36.040306.132643

12. Hessa T, Meindl-Beinker NM, Bernsel A, Kim H, Sato Y, Lerch-Bader M, Nilsson I, White SH, von Heijne G (2007) Molecular code for transmembrane-helix recognition by the Sec61 translocon. Nature 450(7172):1026–1030. doi:10.1038/nature06387

13. Sharpe HJ, Stevens TJ, Munro S (2010) A comprehensive comparison of transmembrane domains reveals organelle-specific properties. Cell 142(1):158–169. doi:10.1016/j.cell.2010.05.037

14. Killian JA (1998) Hydrophobic mismatch between proteins and lipids in membranes. Biochim Biophys Acta 1376(3):401–415

15. Killian JA, von Heijne G (2000) How proteins adapt to a membrane-water interface. Trends Biochem Sci 25(9):429–434

16. Caputo GA, London E (2003) Cumulative effects of amino acid substitutions and hydrophobic mismatch upon the transmembrane stability and conformation of hydrophobic alpha-helices. Biochemistry 42(11):3275–3285. doi:10.1021/bi026697d

17. Bretscher MS, Munro S (1993) Cholesterol and the Golgi apparatus. Science 261(5126): 1280–1281

18. Perozo E, Kloda A, Cortes DM, Martinac B (2002) Physical principles underlying the transduction of bilayer deformation forces during mechanosensitive channel gating. Nat Struct Biol 9(9):696–703. doi:10.1038/nsb827

19. Kloda A, Petrov E, Meyer GR, Nguyen T, Hurst AC, Hool L, Martinac B (2008) Mechanosensitive channel of large conductance. Int J Biochem Cell Biol 40(2):164–169. doi:10.1016/j.biocel.2007.02.003

20. de Planque MR, Greathouse DV, Koeppe RE II, Schafer H, Marsh D, Killian JA (1998) Influence of lipid/peptide hydrophobic mismatch on the thickness of diacylphosphatidylcholine bilayers. A 2H NMR and ESR study using designed transmembrane alpha-helical peptides and gramicidin A. Biochemistry 37(26):9333–9345. doi:10.1021/bi980233r

21. Ramadurai S, Holt A, Schafer LV, Krasnikov VV, Rijkers DT, Marrink SJ, Killian JA, Poolman B (2010) Influence of hydrophobic mismatch and amino acid composition on the lateral diffusion of transmembrane peptides. Biophys J 99(5):1447–1454. doi:10.1016/j.bpj.2010.05.042

22. van der Wel PC, Pott T, Morein S, Greathouse DV, Koeppe RE II, Killian JA (2000) Tryptophan-anchored transmembrane peptides promote formation of nonlamellar phases in phosphatidylethanolamine model membranes in a mismatch-dependent manner. Biochemistry 39(11):3124–3133

23. Zhang YP, Lewis RN, Hodges RS, McElhaney RN (1992) Interaction of a peptide model of a hydrophobic transmembrane alpha-helical segment of a membrane protein with phosphatidylcholine bilayers: differential scanning calorimetric and FTIR spectroscopic studies. Biochemistry 31(46):11579–11588

24. Krishnakumar SS, London E (2007) Effect of sequence hydrophobicity and bilayer width upon the minimum length required for the formation of transmembrane helices in membranes. J Mol Biol 374(3):671–687

25. Kim T, Im W (2010) Revisiting hydrophobic mismatch with free energy simulation studies of transmembrane helix tilt and rotation. Biophys J 99(1):175–183. doi:10.1016/j.bpj.2010.04.015

26. Webb RJ, East JM, Sharma RP, Lee AG (1998) Hydrophobic mismatch and the incorporation of peptides into lipid bilayers: a possible mechanism for retention in the Golgi. Biochemistry 37(2):673–679. doi:10.1021/bi972441+

27. Weiss TM, van der Wel PC, Killian JA, Koeppe RE II, Huang HW (2003) Hydrophobic mismatch between helices and lipid bilayers. Biophys J 84(1):379–385. doi:10.1016/S0006-3495(03)74858-9

28. Morein S, Killian JA, Sperotto MM (2002) Characterization of the thermotropic behavior and lateral organization of lipid-peptide mixtures by a combined experimental and theoretical approach: effects of hydrophobic mismatch and role of flanking residues. Biophys J 82(3):1405–1417. doi:10.1016/S0006-3495(02)75495-7

29. Morein S, Koeppe IR, Lindblom G, de Kruijff B, Killian JA (2000) The effect of peptide/lipid hydrophobic mismatch on the phase behavior of model membranes mimicking the lipid composition in Escherichia coli membranes. Biophys J 78(5):2475–2485

30. Kandasamy SK, Larson RG (2006) Molecular dynamics simulations of model transmembrane peptides in lipid bilayers: a systematic investigation of hydrophobic mismatch. Biophys J 90(7):2326–2343. doi:10.1529/biophysj.105.073395

31. Holt A, Rougier L, Reat V, Jolibois F, Saurel O, Czaplicki J, Killian JA, Milon A (2010) Order parameters of a transmembrane helix in a fluid bilayer: case study of a WALP peptide. Biophys J 98(9):1864–1872. doi:10.1016/j.bpj.2010.01.016

32. Strandberg E, Morein S, Rijkers DT, Liskamp RM, van der Wel PC, Killian JA (2002) Lipid dependence of membrane anchoring properties and snorkeling behavior of aromatic and charged residues in transmembrane peptides. Biochemistry 41(23):7190–7198

33. Liu F, Lewis RN, Hodges RS, McElhaney RN (2002) Effect of variations in the structure of a polyleucine-based alpha-helical transmembrane peptide on its interaction with phosphatidylcholine bilayers. Biochemistry 41(29):9197–9207

34. Liu F, Lewis RN, Hodges RS, McElhaney RN (2004) Effect of variations in the structure of a

polyleucine-based alpha-helical transmembrane peptide on its interaction with phosphatidylethanolamine Bilayers. Biophys J 87(4):2470–2482. doi:10.1529/biophysj.104.046342

35. Liu F, Lewis RN, Hodges RS, McElhaney RN (2004) Effect of variations in the structure of a polyleucine-based alpha-helical transmembrane peptide on its interaction with phosphatidylglycerol bilayers. Biochemistry 43(12): 3679–3687. doi:10.1021/bi036214l

36. Zhang YP, Lewis RN, Hodges RS, McElhaney RN (1995) Interaction of a peptide model of a hydrophobic transmembrane alpha-helical segment of a membrane protein with phosphatidylethanolamine bilayers: differential scanning calorimetric and Fourier transform infrared spectroscopic studies. Biophys J 68(3):847–857. doi:10.1016/S0006-3495(95)80261-4

37. Zhang YP, Lewis RN, Hodges RS, McElhaney RN (2001) Peptide models of the helical hydrophobic transmembrane segments of membrane proteins: interactions of acetyl-K2-(LA)12-K2-amide with phosphatidylethanolamine bilayer membranes. Biochemistry 40(2):474–482

38. Caputo GA, London E (2004) Position and ionization state of Asp in the core of membrane-inserted alpha helices control both the equilibrium between transmembrane and nontransmembrane helix topography and transmembrane helix positioning. Biochemistry 43(27):8794–8806

39. Ren J, Lew S, Wang J, London E (1999) Control of the transmembrane orientation and interhelical interactions within membranes by hydrophobic helix length. Biochemistry 38(18):5905–5912

40. Ren J, Lew S, Wang Z, London E (1997) Transmembrane orientation of hydrophobic alpha-helices is regulated both by the relationship of helix length to bilayer thickness and by the cholesterol concentration. Biochemistry 36(33):10213–10220

41. Krishnakumar SS, London E (2007) The control of transmembrane helix transverse position in membranes by hydrophilic residues. J Mol Biol 374(5):1251–1269. doi:10.1016/j.jmb.2007.10.032

42. Ipsen JH, Mouritsen OG, Bloom M (1990) Relationships between lipid membrane area, hydrophobic thickness, and acyl-chain orientational order. The effects of cholesterol. Biophys J 57(3):405–412. doi:10.1016/S0006-3495(90)82557-1

43. King GI, White SH (1986) Determining bilayer hydrocarbon thickness from neutron diffraction measurements using strip-function models. Biophys J 49(5):1047–1054. doi:10.1016/S0006-3495(86)83733-X

44. Lewis BA, Engelman DM (1983) Lipid bilayer thickness varies linearly with acyl chain length in fluid phosphatidylcholine vesicles. J Mol Biol 166(2):211–217

45. Rawicz W, Olbrich KC, McIntosh T, Needham D, Evans E (2000) Effect of chain length and unsaturation on elasticity of lipid bilayers. Biophys J 79(1):328–339. doi:10.1016/S0006-3495(00)76295-3

46. Caputo GA, London E (2003) Using a novel dual fluorescence quenching assay for measurement of tryptophan depth within lipid bilayers to determine hydrophobic alpha-helix locations within membranes. Biochemistry 42(11):3265–3274

47. Kauko A, Hedin LE, Thebaud E, Cristobal S, Elofsson A, von Heijne G (2010) Repositioning of transmembrane alpha-helices during membrane protein folding. J Mol Biol 397(1):190–201. doi:10.1016/j.jmb.2010.01.042

48. Hessa T, Monne M, von Heijne G (2003) Stop-transfer efficiency of marginally hydrophobic segments depends on the length of the carboxy-terminal tail. EMBO Rep 4(2):178–183. doi:10.1038/sj.embor.embor728

49. Monne M, Nilsson I, Elofsson A, von Heijne G (1999) Turns in transmembrane helices: determination of the minimal length of a "helical hairpin" and derivation of a fine-grained turn propensity scale. J Mol Biol 293(4):807–814. doi:10.1006/jmbi.1999.3183

50. Monne M, Nilsson I, Johansson M, Elmhed N, von Heijne G (1998) Positively and negatively charged residues have different effects on the position in the membrane of a model transmembrane helix. J Mol Biol 284(4):1177–1183. doi:10.1006/jmbi.1998.2218

51. Nilsson I, Saaf A, Whitley P, Gafvelin G, Waller C, von Heijne G (1998) Proline-induced disruption of a transmembrane alpha-helix in its natural environment. J Mol Biol 284(4):1165–1175. doi:10.1006/jmbi.1998.2217

52. Sperotto MM, Mouritsen OG (1991) Monte Carlo simulation studies of lipid order parameter profiles near integral membrane proteins. Biophys J 59(2):261–270. doi:10.1016/S0006-3495(91)82219-6

53. Harroun TA, Heller WT, Weiss TM, Yang L, Huang HW (1999) Theoretical analysis of hydrophobic matching and membrane-mediated interactions in lipid bilayers contain-

ing gramicidin. Biophys J 76(6):3176–3185. doi:10.1016/S0006-3495(99)77469-2

54. Harroun TA, Heller WT, Weiss TM, Yang L, Huang HW (1999) Experimental evidence for hydrophobic matching and membrane-mediated interactions in lipid bilayers containing gramicidin. Biophys J 76(2):937–945. doi:10.1016/S0006-3495(99)77257-7

55. Liu F, Lewis RN, Hodges RS, McElhaney RN (2001) A differential scanning calorimetric and 31P NMR spectroscopic study of the effect of transmembrane alpha-helical peptides on the lamellar-reversed hexagonal phase transition of phosphatidylethanolamine model membranes. Biochemistry 40(3):760–768

56. Harzer U, Bechinger B (2000) Alignment of lysine-anchored membrane peptides under conditions of hydrophobic mismatch: a CD, 15N and 31P solid-state NMR spectroscopy investigation. Biochemistry 39(43):13106–13114

57. Duong-Ly KC, Nanda V, Degrado WF, Howard KP (2005) The conformation of the pore region of the M2 proton channel depends on lipid bilayer environment. Protein Sci 14(4):856–861. doi:10.1110/ps.041185805

58. Nguyen PA, Soto CS, Polishchuk A, Caputo GA, Tatko CD, Ma C, Ohigashi Y, Pinto LH, DeGrado WF, Howard KP (2008) pH-induced conformational change of the influenza M2 protein C-terminal domain. Biochemistry 47(38):9934–9936. doi:10.1021/bi801315m

59. Strandberg E, Ozdirekcan S, Rijkers DT, van der Wel PC, Koeppe RE II, Liskamp RM, Killian JA (2004) Tilt angles of transmembrane model peptides in oriented and non-oriented lipid bilayers as determined by 2H solid-state NMR. Biophys J 86(6):3709–3721. doi:10.1529/biophysj.103.035402

60. van der Wel PC, Strandberg E, Killian JA, Koeppe RE II (2002) Geometry and intrinsic tilt of a tryptophan-anchored transmembrane alpha-helix determined by (2)H NMR. Biophys J 83(3):1479–1488. doi:10.1016/S0006-3495(02)73918-0

61. de Planque MR, Goormaghtigh E, Greathouse DV, Koeppe RE II, Kruijtzer JA, Liskamp RM, de Kruijff B, Killian JA (2001) Sensitivity of single membrane-spanning alpha-helical peptides to hydrophobic mismatch with a lipid bilayer: effects on backbone structure, orientation, and extent of membrane incorporation. Biochemistry 40(16):5000–5010

62. Burck J, Roth S, Wadhwani P, Afonin S, Kanithasen N, Strandberg E, Ulrich AS (2008) Conformation and membrane orientation of amphiphilic helical peptides by oriented circular dichroism. Biophys J 95(8):3872–3881. doi:10.1529/biophysj.108.136085

63. Clayton AH, Sawyer WH (2000) Oriented circular dichroism of a class A amphipathic helix in aligned phospholipid multilayers. Biochim Biophys Acta 1467(1):124–130

64. Kuball HG, Hofer T (2000) Chirality and circular dichroism of oriented molecules and anisotropic phases. Chirality 12(4):278–286. doi:10.1002/(SICI)1520-636X(2000)12: 4<278::AID-CHIR14>3.0.CO;2-O

65. de Jongh HH, Goormaghtigh E, Killian JA (1994) Analysis of circular dichroism spectra of oriented protein-lipid complexes: toward a general application. Biochemistry 33(48): 14521–14528

66. Wu Y, Huang HW, Olah GA (1990) Method of oriented circular dichroism. Biophys J 57(4):797–806. doi:10.1016/S0006-3495 (90)82599-6

67. Vigano C, Manciu L, Buyse F, Goormaghtigh E, Ruysschaert JM (2000) Attenuated total reflection IR spectroscopy as a tool to investigate the structure, orientation and tertiary structure changes in peptides and membrane proteins. Biopolymers 55(5):373–380. doi:10.1002/1097-0282(2000)55: 5<373::AID-BIP1011>3.0.CO;2-U

68. Reinl HM, Bayerl TM (1993) Interaction of myelin basic protein with single bilayers on a solid support: an NMR, DSC and polarized infrared ATR study. Biochim Biophys Acta 1151(2):127–136

69. Frey S, Tamm LK (1991) Orientation of melittin in phospholipid bilayers. A polarized attenuated total reflection infrared study. Biophys J 60(4):922–930. doi:10.1016/ S0006-3495(91)82126-9

70. Yin H, Litvinov RI, Vilaire G, Zhu H, Li W, Caputo GA, Moore DT, Lear JD, Weisel JW, Degrado WF, Bennett JS (2006) Activation of platelet alphaIIbbeta3 by an exogenous peptide corresponding to the transmembrane domain of alphaIIb. J Biol Chem 281(48):36732–36741. doi:10.1074/jbc.M605877200

71. Yin H, Slusky JS, Berger BW, Walters RS, Vilaire G, Litvinov RI, Lear JD, Caputo GA, Bennett JS, DeGrado WF (2007) Computational design of peptides that target transmembrane helices. Science 315(5820): 1817–1822. doi:10.1126/science.1136782

72. Axelsen PH, Kaufman BK, McElhaney RN, Lewis RN (1995) The infrared dichroism of transmembrane helical polypeptides. Biophys J 69(6):2770–2781. doi:10.1016/ S0006-3495(95)80150-5

73. Ausili A, Corbalan-Garcia S, Gomez-Fernandez JC, Marsh D (2011) Membrane docking of the C2 domain from protein kinase Calpha as seen by polarized ATR-IR. The role of PIP(2).

Biochim Biophys Acta 1808(3):684–695. doi:10.1016/j.bbamem.2010.11.035

74. Lorenz-Fonfria VA, Granell M, Leon X, Leblanc G, Padros E (2009) In-plane and out-of-plane infrared difference spectroscopy unravels tilting of helices and structural changes in a membrane protein upon substrate binding. J Am Chem Soc 131(42):15094–15095. doi:10.1021/ja906324z

75. DeGrado WF, Gratkowski H, Lear JD (2003) How do helix-helix interactions help determine the folds of membrane proteins? Perspectives from the study of homo-oligomeric helical bundles. Protein Sci 12(4): 647–665

76. Lear JD, Stouffer AL, Gratkowski H, Nanda V, Degrado WF (2004) Association of a model transmembrane peptide containing gly in a heptad sequence motif. Biophys J 87(5):3421–3429. doi:10.1529/biophysj.103.032839

77. Runnels LW, Scarlata SF (1995) Theory and application of fluorescence homotransfer to melittin oligomerization. Biophys J 69(4): 1569–1583. doi:10.1016/S0006-3495(95)80030-5

78. You M, Li E, Wimley WC, Hristova K (2005) Forster resonance energy transfer in liposomes: measurements of transmembrane helix dimerization in the native bilayer environment. Anal Biochem 340(1):154–164. doi:10.1016/j.ab.2005.01.035

79. Lew S, Caputo GA, London E (2003) The effect of interactions involving ionizable residues flanking membrane-inserted hydrophobic helices upon helix-helix interaction. Biochemistry 42(36):10833–10842

80. Lew S, London E (1997) Simple procedure for reversed-phase high-performance liquid chromatographic purification of long hydrophobic peptides that form transmembrane helices. Anal Biochem 251(1):113–116

81. Ladokhin AS (1997) Distribution analysis of depth-dependent fluorescence quenching in membranes: a practical guide. Methods Enzymol 278:462–473

82. Abrams FS, London E (1992) Calibration of the parallax fluorescence quenching method for determination of membrane penetration depth: refinement and comparison of quenching by spin-labeled and brominated lipids. Biochemistry 31(23):5312–5322

83. Chattopadhyay A, London E (1987) Parallax method for direct measurement of membrane penetration depth utilizing fluorescence quenching by spin-labeled phospholipids. Biochemistry 26(1):39–45

84. Bolen EJ, Holloway PW (1990) Quenching of tryptophan fluorescence by brominated phospholipid. Biochemistry 29(41):9638–9643

85. Senes A, Chadi DC, Law PB, Walters RF, Nanda V, Degrado WF (2007) E(z), a depth-dependent potential for assessing the energies of insertion of amino acid side-chains into membranes: derivation and applications to determining the orientation of transmembrane and interfacial helices. J Mol Biol 366(2):436–448

86. Schramm CA, Hannigan BT, Donald JE, Keasar C, Saven JG, Degrado WF, Samish I (2012) Knowledge-based potential for positioning membrane-associated structures and assessing residue-specific energetic contributions. Structure 20(5):924–935. doi:10.1016/j.str.2012.03.016

87. Landolt-Marticorena C, Williams KA, Deber CM, Reithmeier RA (1993) Non-random distribution of amino acids in the transmembrane segments of human type I single span membrane proteins. J Mol Biol 229(3): 602–608. doi:10.1006/jmbi.1993.1066

88. Strandberg E, Esteban-Martin S, Ulrich AS, Salgado J (2012) Hydrophobic mismatch of mobile transmembrane helices: merging theory and experiments. Biochim Biophys Acta 1818(5):1242–1249. doi:10.1016/j.bbamem.2012.01.023

89. Holt A, Killian JA (2010) Orientation and dynamics of transmembrane peptides: the power of simple models. Eur Biophys J 39(4): 609–621. doi:10.1007/s00249-009-0567-1

90. Ozdirekcan S, Etchebest C, Killian JA, Fuchs PF (2007) On the orientation of a designed transmembrane peptide: toward the right tilt angle? J Am Chem Soc 129(49): 15174–15181. doi:10.1021/ja073784q

91. Pan J, Tristram-Nagle S, Nagle JF (2009) Alamethicin aggregation in lipid membranes. J Membr Biol 231(1):11–27. doi:10.1007/s00232-009-9199-8

92. Killian JA (2003) Synthetic peptides as models for intrinsic membrane proteins. FEBS Lett 555(1):134–138

93. de Planque MR, Killian JA (2003) Protein-lipid interactions studied with designed transmembrane peptides: role of hydrophobic matching and interfacial anchoring. Mol Membr Biol 20(4):271–284. doi:10.1080/09687680310001605352

94. Rankenberg JM, Vostrikov VV, DuVall CD, Greathouse DV, Koeppe RE II, Grant CV, Opella SJ (2012) Proline kink angle distributions for GWALP23 in lipid bilayers of different thicknesses. Biochemistry 51(17): 3554–3564. doi:10.1021/bi300281k

95. Shahidullah K, London E (2008) Effect of lipid composition on the topography of membrane-associated hydrophobic helices: stabilization of transmembrane topography by anionic lipids. J Mol Biol 379(4):704–718. doi:10.1016/j.jmb.2008.04.026

96. Kremer JM, Esker MW, Pathmamanoharan C, Wiersema PH (1977) Vesicles of variable diameter prepared by a modified injection method. Biochemistry 16(17):3932–3935

97. Aarts PA, Gijeman OL, Kremer JM, Wiersema PH (1977) Dynamics of phospholipid aggregation in ethanol–water solutions. Chem Phys Lipids 19(3):267–274

98. London E (2007) Using model membrane-inserted hydrophobic helices to study the equilibrium between transmembrane and non-transmembrane states. J Gen Physiol 130(2): 229–232

99. Shahidullah K, Krishnakumar SS, London E (2010) The effect of hydrophilic substitutions and anionic lipids upon the transverse positioning of the transmembrane helix of the ErbB2 (neu) protein incorporated into model membrane vesicles. J Mol Biol 396(1): 209–220. doi:10.1016/j.jmb.2009.11.037

100. Jones JD, Gierasch LM (1994) Effect of charged residue substitutions on the membrane-interactive properties of signal sequences of the Escherichia coli LamB protein. Biophys J 67(4):1534–1545. doi:10.1016/ S0006-3495(94)80627-7

101. McIntosh TJ, Simon SA, MacDonald RC (1980) The organization of n-alkanes in lipid bilayers. Biochim Biophys Acta 597(3):445–463

102. Uhrikova D, Balgavy P, Kucerka N, Islamov A, Gordeliy V, Kuklin A (2000) Small-angle neutron scattering study of the n-decane effect on the bilayer thickness in extruded unilamellar dioleoylphosphatidylcholine liposomes. Biophys Chem 88(1–3): 165–170

103. Nezil FA, Bloom M (1992) Combined influence of cholesterol and synthetic amphiphillic peptides upon bilayer thickness in model membranes. Biophys J 61(5):1176–1183. doi:10.1016/S0006-3495(92)81926-4

104. Sanchez SA, Gunther G, Tricerri MA, Gratton E (2011) Methyl-beta-cyclodextrins preferentially remove cholesterol from the liquid disordered phase in giant unilamellar vesicles. J Membr Biol 241(1):1–10. doi:10.1007/ s00232-011-9348-8

105. Veatch SL, Keller SL (2003) Separation of liquid phases in giant vesicles of ternary mixtures of phospholipids and cholesterol. Biophys J 85(5):3074–3083. doi:10.1016/ S0006-3495(03)74726-2

106. Caputo GA, Litvinov RI, Li W, Bennett JS, Degrado WF, Yin H (2008) Computationally designed peptide inhibitors of protein-protein interactions in membranes. Biochemistry 47(33):8600–8606. doi:10.1021/ bi800687h

107. Hayashibara M, London E (2005) Topography of diphtheria toxin A chain inserted into lipid vesicles. Biochemistry 44(6): 2183–2196

108. Follenius-Wund A, Piemont E, Freyssinet JM, Gerard D, Pigault C (1997) Conformational adaptation of annexin V upon binding to liposomes: a time-resolved fluorescence study. Biochem Biophys Res Commun 234(1):111–116. doi:10.1006/ bbrc.1997.6596

Chapter 6

Folding Alpha-Helical Membrane Proteins into Liposomes In Vitro and Determination of Secondary Structure

Heather E. Findlay and Paula J. Booth

Abstract

The native environment of integral membrane proteins is a highly complex lipid bilayer composed of many different types of lipids, the physical characteristics of which can profoundly influence protein stability, folding, and function. Secondary transporters are a class of protein where changes to both structure and activity have been observed in different bilayer environments. In order to study these interactions in vitro, it is necessary to extract and purify the protein and exchange it into an artificial lipid system that can be manipulated to control protein behavior. Liposomes are a commonly used model system that is particularly suitable for studying transporters. GalP and LacY can be reconstituted or refolded into vesicles with a high degree of efficiency for further structural analysis. Circular dichroism spectroscopy is an important technique in monitoring protein folding, which allows the decomposition of spectra into secondary structural components.

Key words Membrane protein, Folding, Circular dichroism, Lipid vesicle, Flotation assay

1 Introduction

The folding and assembly of a protein from its amino acid sequence to its final 3D structure is a fundamental cellular process, but the general principles of protein folding have to date largely emerged from small water-soluble proteins. For membrane proteins the picture is more complicated, with the surrounding bilayer environment—comprises lipids with a variety of headgroup, charge, and aliphatic chains—profoundly influencing the folding process as well as protein stability and function [1, 2]. This heterogeneric membrane environment is not amenable to many of the biophysical, aqueous-focussed techniques commonly used to study proteins. Therefore membrane proteins are typically extracted from their bilayer and reconstituted into a model system that can act as a mimic of their native membrane.

Giovanna Ghirlanda and Alessandro Senes (eds.), *Membrane Proteins: Folding, Association, and Design*,
Methods in Molecular Biology, vol. 1063, DOI 10.1007/978-1-62703-583-5_6, © Springer Science+Business Media, LLC 2013

One commonly used model is the liposome or lipid vesicle. Liposomes are spherical vesicles with an internal aqueous compartment surrounded by a single lipid bilayer. They range in size from "small" and "large" unilamellar vesicles with diameters of tens to hundreds of nm to giant vesicles up to 100 μm across. There are several advantages of working with liposomes. They provide a good bilayer mimic and can be formed from a broad range of lipids, with mixes of both lamellar and non-lamellar lipids, allowing the study of both specific and global lipid effects. They are a particularly good system for studying transport proteins as the isolated internal compartment provides the directionality required for activity assays.

Membrane proteins are generally isolated from their native membranes via detergent extraction and then reconstituted back into a lipid bilayer. This reconstitution of detergent-solubilized protein into liposomes rarely occurs with 100 % efficiency, so it is often necessary to separate misfolded or non-incorporated protein from that associated with the vesicles. Vesicles are typically removed from solution by high-speed centrifugation, but the same conditions can also sediment aggregated protein. A useful method that gives a much better separation is a sucrose flotation assay. Here, liposomes resuspended in a concentrated sucrose solution are layered at the bottom of a sucrose gradient. After centrifugation, misfolded and aggregated protein is left at the bottom of the tube, while the liposomes and their associated protein move up to the top of the gradient. During reconstitution membrane proteins are not intentionally unfolded; the aim is to maintain the functional fold during detergent extraction and the subsequent reconstitution. In contrast, some membrane proteins have been successfully folded directly into liposomes from partly denatured states. These denatured states involve unfolding the protein (generally from its folded, stable detergent-extracted state) in denaturing agents such as urea, guanidinium hydrochloride, or sodium dodecyl sulfate. Helical membrane proteins do not lose all their secondary structure in these denaturants but do lose key tertiary structural elements together with loss of function. In certain cases membrane proteins have been spontaneously refolded into liposomes directly from these partly denatured states merely by diluting the denaturant. A flotation assay is very useful in removing aggregated and misfolded protein.

Another vital technique for the study of protein folding and reconstitution is circular dichroism (CD). This spectroscopic method involves the differential absorption of left- and right-handed circularly polarized light by the sample. In the far-UV region below 260 nm elements of secondary structure display characteristic spectra, which enables the analysis of a protein spectrum to reveal the contributing secondary structural features [3, 4]. Far-UV CD therefore enables changes in secondary structure to be followed during unfolding and refolding experiments.

The galactose and lactose transporters GalP and LacY are secondary transporters from *Escherichia coli* that drive sugar import into the cell utilizing a proton gradient. They are members of the major facilitator superfamily (MFS), a large group of alpha-helical membrane proteins that are estimated to comprise 25 % of all known transport proteins in prokaryotes [5]. MFS proteins have a common fold of two 6-helical bundles with the substrate binding site between these domains and are proposed to achieve transport by a mechanism of alternating access to opposing sides of the membrane. The lipid environment surrounding these proteins is known to influence their structure and function [6, 7].

2 Materials

All chemicals are purchased from Sigma unless otherwise stated.

2.1 Preparation of Liposomes, Protein Reconstitution, and Folding

1. 1,2-dioleoyl-sn-glycero-3-phosphocholine (DOPC), 1,2-dioleoyl-sn-glycero-3-phosphoethanolamine (DOPE), and 1,2-dioleoyl-sn-glycero-3-[phospho-rac-(1-glycerol)] (DOPG) as powder and 1,2-dioleoyl-sn-glycero-3-phosphoethanolamine-*N*-(lissamine rhodamine B sulfonyl) at 1 mg/mL in chloroform from Avanti Polar Lipids.

2. 50 mM sodium phosphate pH8 buffer (NaPhos buffer).

3. 100 nm polycarbonate filters (Nucleopore Track-Etch Membrane) from Whatman and Mini-Extruder kit from Avanti Polar Lipids.

4. Octyl-b-D-glucoside (OG) and dodecyl-b-D-maltoside (DDM) from Anatrace.

5. Purified GalP or LacY in NaPhos buffer including 0.05 % (w/v) DDM at a concentration of 4 mg/mL or 2 mg/mL, respectively.

6. Biobeads from BioRad washed with methanol and water, as per the manufacturer's instructions, then presoaked in NaPhos buffer.

7. Unfolding buffer—10 M urea in 50 mM sodium phosphate pH8.

2.2 Sucrose Flotation Assay

1. 60 and 15 % (w/v) sucrose dissolved in 50 mM sodium phosphate pH 8.

2. Beckman TLS55 swing-out rotor and 2.2 mL thinwall polyallomer tubes.

2.3 Circular Dichroism Spectroscopy

1. Synchrotron Radiation Source (SRS) beamline or CD spectrophotometer with modified detector geometry to capture scattered light. Aviv Model 410 CD-spectrometer can be optimized in this way.

2. Strain-free quartz Suprasil demountable cuvettes from Hellma.

3   Methods

3.1   Preparation of Liposomes and Reconstitution of GalP

1. Prepare stock solutions of DOPC, DOPE, and DOPG at 50 mg/mL in cyclohexane and mix to the required molar ratio in small glass vials. If desired, add 0.1 mol % rhodamine-labelled DOPE to allow tracking of the vesicles spectroscopically. Freeze the lipids over liquid nitrogen and evaporate the solvent overnight under vacuum.

2. Resuspend the lipid film in NaPhos buffer at 10 mg/mL lipid and mix at room temperature for 15 min, vortexing occasionally. Extrude the liposomes to 100 nm diameter by passing 11 times through a polycarbonate filter using an Avanti Mini-Extruder.

3. Add OG from a 20 % (w/v) stock solution to the liposomes to a final concentration 1 % to pre-saturate the vesicles and incubate for 10 min. Add the purified protein to a final concentration of 0.2 mg/mL (w/v) and incubate for a further 30 min.

4. Excess detergent is removed by incubating the proteoliposomes with approximately 200 mg wet beads per mL vesicles for 1 h. The beads are spun down in a benchtop microcentrifuge, the supernatant retained and fresh beads added. This step is repeated twice, resulting in detergent-free reconstituted protein (see **Note 1**).

3.2   Folding of Denatured GalP into Liposomes

1. Prepare 100 nm liposomes as described for reconstitution **steps 1** and **2**.

2. Mix purified GalP 1 in 5 in 10 M urea in NaPhos buffer to give a final concentration of 8 M urea. Incubate for 2 min at room temperature to unfold the protein (see **Note 2**).

3. Dilute the denatured protein tenfold into 10 mg/mL liposomes in NaPhos buffer and incubate for a further 10 min to allow the protein to refold. Residual urea can be removed by dialysis against NaPhos buffer.

3.3   Sucrose Flotation Assay

1. Add 200 μL of 60 % sucrose solution to 200 μL of prepared proteoliposomes. Pipette into the bottom of a 2.2 mL volume thinwall polyallomer tube.

2. Carefully layer 1.6 mL of 15 % sucrose solution over the proteoliposomes. The two layers should be clearly visible with negligible mixing of the two.

3. Carefully layer 0.2 mL of NaPhos buffer over the 15 % sucrose layer. Again the layers should be distinct (see **Note 3**).

4. Load the tubes into a Beckman TLS55 swing-out rotor. Centrifuge for 1 h at 55,000 rpm (200,000×g). After centrifugation the liposomes and associated protein will have floated to the top of the sucrose gradient. Remove and retain the top layer (Fig. 1a).

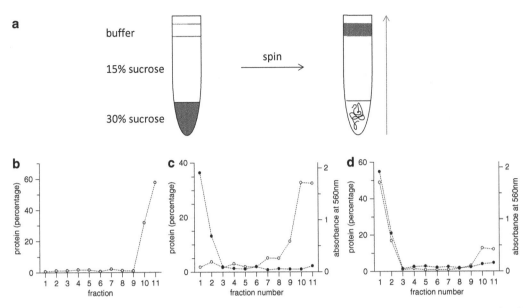

Fig. 1 Sucrose flotation assay. (**a**) Schematic. Liposomes containing 0.1 % rhodamine-DOPE were mixed with sucrose to a final concentration of 30 % (w/v). 15 % (w/v) sucrose solution and buffer were layered on top. After centrifugation, liposomes floated up the gradient. (**b**) LacY was added to the bottom of a gradient without liposomes present. After the flotation 200 μL fractions were taken (1–11, *top* to *bottom*) and analyzed for protein content. (**c**) Bovine serum albumin was mixed with liposomes, added to a gradient, and centrifuged. Fractions were analyzed for protein content (*open circles*) and lipid content (*closed circles*). (**d**) LacY was reconstituted into liposomes and added to a gradient. After centrifugation fractions were analyzed for protein content (*open circles*) and lipid content (*closed circles*)

5. If rhodamine-DOPE was included in the lipid mix, the liposomes can be detected by measuring the absorbance at 560 nm down the sucrose gradient (Fig. 1c, d).

6. The protein can be detected and its concentration determined either by SDS-PAGE followed by densitometric analysis or using a lipid-independent concentration assay (e.g., Markwell-Lowry assay [8] or amido black assay [9], *see* **Note 4**) (Fig. 1b–d).

3.4 Circular Dichroism Spectroscopy

1. If possible circular dichroism measurements should be recorded at an SRS, where the high signal intensity increases signal to noise at the low UV wavelengths where the lipids absorb and thus high-quality spectra can be gained down to ~180 nm. Spectra can also be measured successfully with an Aviv Model 410 CD-spectrometer adapted for samples that scatter light (*see* **Note 5**).

2. Load the proteoliposome sample into a 0.1 mm round demountable quartz cuvette. Record spectra from 260 to 180 nm with a data pitch of 1 nm and a dwell time of 1 s. If using a non-SRS machine, increase the dwell time to 3 s. Scan 4 times and average the traces to reduce the noise in the

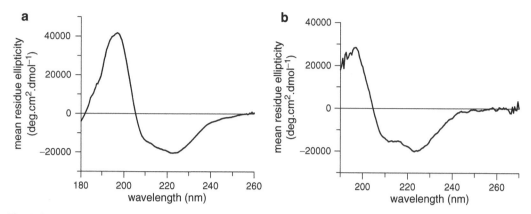

Fig. 2 Circular dichroism spectroscopy. (**a**) LacY was reconstituted into liposomes composed of 40 % DOPC, 40%DOPE, and 20%DOPG. A spectrum was measured from 260 to 180 nm at a synchrotron radiation source beamline. (**b**) GalP was unfolded in 8 M urea then refolded into liposomes composed of 40 % DOPC and 60 % DOPE. A CD spectrum was measured from 270 to 190 nm

spectrum. Make similar measurements of liposomes prepared without added protein and subtract from the proteoliposome sample (Fig. 2).

3. When the data extends into the far UV until at least 190 nm, it is possible to decompose the spectra into the secondary structure components. This is carried using the analysis program CDSSTR at the DICHROWEB website, in which the spectrum is fitted using a series of proteins of known structure as a reference dataset (*see* **Note 6**).

4 Notes

1. During the reconstitution process detergent is added both to pre-saturate the liposome and as part of the detergent-extracted protein sample. It is important that this detergent is removed after reconstitution so that the bilayer remains intact and non-leaky in further experiments including the flotation assay. When the detergent is removed, the samples will no longer contain bubbles. The exact amount of residual detergent can be measured using a phenol/sulfuric concentration assay [10] and should be reduced to 0.01 % (w/v) or better.

2. Urea is a chaotropic agent commonly used in folding studies for the denaturation of proteins. In solution urea degrades into isocyanic acid, more rapidly at alkaline pH, which will irreversibly carbamylate proteins at amino groups. This process can significantly affect both the unfolding and even more so the refolding properties. For this reason, it is essential that urea solutions are freshly made before every experiment.

GalP does not refold in detergent micelles if the urea is not fresh. Urea solutions can also be deionized by passing them over an ion-exchange resin.

3. Good separation of vesicles from protein in the flotation assay relies on both the quality and homogeneity of sample and well-distinguished sucrose layers. Liposomes prepared by extrusion should result in a tight distribution of size around that of the filter pore. If necessary this can be confirmed by dynamic light scattering. Sucrose will diffuse into leaky vesicles, causing them to be retained in the lower sucrose layers. If this is a problem the reconstitution procedure may need to be optimized. Adjusting the protein to lipid ratio, minimizing residual detergent, and changing temperature will all affect the bilayer leakage. Preparing the sucrose gradients can take practice and a steady hand. Rather than using a pipette, it is often easier to layer the sucrose using a syringe with a wide-blunted needle.

4. To analyze circular dichroism data, it is important to have an accurate measure of protein concentration. Many assays cannot be used in the presence of detergents or lipids. The Markwell-Lowry assay and the amido black assay both involve a precipitation step in trichloroacetic acid to wash these away. An alternative method is to dissolve the liposomes in organic solvent such as chloroform:methanol and measuring the absorbance at 280 nm.

5. Light scattering is a well-known problem for spectroscopic experiments on membrane protein samples. Circular dichroism spectra are distorted by differential scattering and show flattening of the absorption peaks. It is important to minimize this for accurate measurements. Firstly the sample size should not be too large, with liposomes prepared to no greater than 100 nm if possible. Secondly the machine should be adapted to collect light from increased angles. Modified CD spectrophotometers have detectors with high collection angles, set close to the sample cell to maximize gathered light. Synchrotron light sources are typically set up in this way as standard.

6. Good-quality spectra are required for accurate assignation of secondary structure. The dynode signal from the spectrophotometer provides a useful guide. As the sample total absorbance increases, the dynode also goes up to compensate. Once the signal rises above 600 V, the CD data is no longer reliable and should be cut off. The CD spectrum must cover the range of 260–190 nm at least to allow deconvolution into a-helix, b-sheet, turn and unordered using the software available through DICHROWEB, which compares the spectrum to a reference set of proteins of known structure. If the spectrum extends to 180 nm, the SMP180 reference set can be used [11]. In contrast to the others, this reference set also contains

membrane proteins and therefore gives an improved structural prediction. One important way to improve spectra is to optimize the buffer for CD, so that any chiral compounds are excluded and the absorbance of the solution in the far UV is minimized. Good's buffers are poor for CD experiments, phosphate buffers are good, and Tris intermediate. Chloride ions should be avoided if possible. If not possible, the salt can often be substituted for the fluoride ion equivalent.

References

1. Bowie JU (2005) Solving the membrane protein folding problem. Nature 438(7068): 581–589
2. Findlay HE, Booth PJ (2006) The biological significance of lipid-protein interactions. J Phys Condens Matter 18(28):S1281–S1291
3. Whitmore L, Wallace BA (2008) Protein secondary structure analyses from circular dichroism spectroscopy: methods and reference databases. Biopolymers 89(5):392–400
4. Whitmore L, Wallace BA (2004) DICHROWEB, an online server for protein secondary structure analyses from circular dichroism spectroscopic data. Nucleic Acids Res 32(web server issue):W668–W673
5. Law CJ, Maloney PC, Wang DN (2008) Ins and outs of major facilitator superfamily antiporters. Annu Rev Microbiol 62:289–305
6. Bogdanov M, Heacock PN, Dowhan W (2002) A polytopic membrane protein displays a reversible topology dependent on membrane lipid composition. EMBO J 21(9): 2107–2116
7. Findlay HE, Rutherford NG, Henderson PJ, Booth PJ (2010) Unfolding free energy of a two-domain transmembrane sugar transport protein. Proc Natl Acad Sci USA 107(43): 18451–18456
8. Markwell MAK, Haas SM, Bieber LL, Tolbert NE (1978) Modification of lowry procedure to simplify protein determination in membrane and lipoprotein samples. Anal Biochem 87(1):206–210
9. Schaffner W, Weissmann C (1973) A rapid, sensitive, and specific method for the determination of protein in dilute solution. Anal Biochem 56(2):502–514
10. Dubois M, Gilles KA, Hamilton JK, Rebers PA, Smith F (1956) Colorimetric method for determination of sugars and related substances. Anal Chem 28(3):350–356
11. Abdul-Gader A, Miles AJ, Wallace BA (2011) A reference dataset for the analyses of membrane protein secondary structures and transmembrane residues using circular dichroism spectroscopy. Bioinformatics 27(12):1630–1636

Chapter 7

Solvation Models and Computational Prediction of Orientations of Peptides and Proteins in Membranes

Andrei L. Lomize and Irina D. Pogozheva

Abstract

Membrane-associated peptides and proteins function in the highly heterogeneous environment of the lipid bilayer whose physico-chemical properties change non-monotonically along the bilayer normal. To simulate insertion of peptides and proteins into membranes and correctly reproduce the energetics of this process, an adequate solvation model and physically realistic representation of the lipid bilayer should be employed. We present a brief overview of the existing solvation models and their application for prediction of binding affinities and orientations of proteins in membranes. Particular emphasis is placed on the recently proposed PPM method, the corresponding web server, and the OPM database that were designed for positioning in membranes of integral and peripheral proteins with known three-dimensional structures.

Key words Transmembrane proteins, Peripheral proteins, Database, Web server, Solvent-accessible surface area

1 Introduction

Development of reliable, accurate, and efficient methods for modeling of partitioning and spatial arrangement of peptides and proteins in biological membranes is an important problem in computational chemistry. The biological membranes provide a highly anisotropic environment with a sharp transition from the water-rich lipid head group region to the hydrophobic acyl chain interior. Immersion into the lipid bilayer restricts the mutual arrangement of interacting peripheral and transmembrane (TM) proteins and affects the strength of all intermolecular interactions that are known to depend on polarity of the environment: the hydrophobic effect, hydrogen bonds, electrostatic forces, and van der Waals (vdW) interactions [1, 2]. Therefore, the spatial arrangement of proteins in membranes is a significant factor that influences protein structure, thermodynamic stability, biological functions, and oligomerization.

Giovanna Ghirlanda and Alessandro Senes (eds.), *Membrane Proteins: Folding, Association, and Design*,
Methods in Molecular Biology, vol. 1063, DOI 10.1007/978-1-62703-583-5_7, © Springer Science+Business Media, LLC 2013

The ongoing progress in methods of protein structure determination accelerates the growth of the number of known three-dimensional (3D) structures of TM and peripheral proteins, which currently constitute around 2 % and 13 % of all entries in the Protein Data Bank (PDB) [3], respectively [4]. However, the location of membrane boundaries is not obvious from these structures, even if a few co-crystallized lipids are present. Experimental techniques for determination of protein arrangement in membranes are too laborious to keep up with the increasing flow of solved experimental structures and, hence, should be supplemented by reliable computational methods.

A number of computational methods can be used for positioning of proteins with known 3D structures in membranes [5]. To define the optimal spatial location of peptides and proteins in membranes, it is essential to accurately reproduce their transfer free energies from water to the lipid bilayer. These energies can be experimentally determined by measuring the corresponding partition coefficients [6]. The theoretical estimates must be sufficiently precise because membrane-binding energies are relatively small for peptides and peripheral proteins [7]. To achieve this goal, it is essential to have an adequate solvation model and a proper physical description of the lipid bilayer as an anisotropic media. The solvation model must account for all main components of the transfer energy, including hydrophobic interactions, solute–solvent hydrogen bonds, and electrostatic interactions of protein with water and lipids. The membrane representation should describe the non-monotonous changes of its chemical and physical properties which depend on the lipid composition [8].

All-atom molecular dynamics (MD) or Monte Carlo (MC) simulations of proteins in the explicit lipid bilayer provide the highest level of molecular details [9]. Development of molecular mechanics (MM) force fields [10–12] and improvement of parameters for lipids (see Lipidbook database [13]) help to better reproduce experimental measurements. However, despite advances in methodology and technology, unconstrained MD simulations applied for protein structure refinement fail to find the correct global minimum of free energy. Indeed, it was recently demonstrated that during sufficiently long (at least 100 μs) all-atom MD simulations crystal structures and the majority of homology models tested (20 out of 24) drifted away from the native structure [14]. These findings point out some deficiencies in the underlying force fields [14, 15]. In particular, the standard MM force fields neglect the solvent and solute polarization [16, 17] and the environment-dependence of vdW forces [1]. Moreover, MD and MC simulations optimize the potential energy or enthalpy of the system [18], but ignore conformational entropy and the entropic part of the hydrophobic effect, two major contributions to stability and solvation of proteins. The calculation of free energy requires an extensive conformational sampling to estimate statistical parameters of the system, including its entropy [19].

Small errors in the force fields accumulate during calculations of large potential energies and conformational sampling. Hence, the MD simulations are expected to be less accurate than advanced continuum solvent models [20, 21] whose parameters are derived from experimental partition coefficients and, therefore, directly convey the required solvation free energy.

In addition, all-atom MD simulations in explicit solvent are rather sophisticated, difficult to use by nonexperts, and computationally highly expensive. These simulations remain practical for relatively simple systems (up to ~100,000 atoms) and short processes (up to ~100 ns), such as insertion of small peptides in one-component or multicomponent lipid bilayer [22, 23].

Modeling of processes at larger time and size scales is achievable by less accurate methods, such as coarse-grained (CG) MD simulations, where groups of interacting atoms of proteins and lipids are treated as large particles [24]. Recent CGMD simulations with MARTINI force field [16, 25] have been successfully applied for positioning in membranes of a large number of membrane peptides and proteins [26–31]. The results of these simulations for more than a hundred TM proteins and several peripheral proteins were collected in a specialized CG database [30]. A more laborious but more informative approach for studying large protein–lipid systems employs multi-scale simulations, where CGMD is combined with subsequent conversion to atomic resolution [5, 32]. The produced models were shown to be accurate enough to compare with experimental data [26]. Main problems of the CG approach are related to parameterization of corresponding force fields, which are not transferrable for different particle configurations, and verification of the method against a sufficiently large set of experimental data [24]. The CG simulations do not optimize directly the required free energy of protein–membrane association, similar to other MD techniques. The free energy of helix insertion into membrane was obtained indirectly from the potential of mean force derived from CGMD simulations, and it appeared to be much greater than that expected from translocon-mediated insertion [29].

Another approach to increase the computational efficiency of MD simulations is to use implicit solvent models of the lipid bilayer in combination with all-atom representation of proteins [33]. Such methods treat the lipid bilayer as a low-dielectric slab embedded between two aqueous regions with continuous dielectric boundaries, which can be described by linear [34] or sigmoidal smoothing functions [35]. To deal with heterogeneous dielectric environment, some of these methods represent the membrane as a series of dielectric slabs [36] or use position-dependent scaling factors [37, 38]. However, choice of smoothing functions, slab thicknesses, and their dielectric constants is rather arbitrary and does not depend on lipid composition.

Implicit solvent models usually describe the solvation energy as a sum of electrostatic and hydrophobic components [39]. To account for energy of electrostatic interactions, the finite difference solution of the Poisson equation [40] or generalized Born (GB) method [41] can be employed. The non-electrostatic energy contribution is usually calculated as a product of ASA of the molecule and a surface tension coefficient describing the energy of cavity formation which is identical for all types of atoms. These methods quite successfully estimate the long-range electrostatic component of solvation energy. However, they assume that polarity of a solvent can be characterized solely by the macroscopic dielectric constant without taking into account the hydrogen-bonding donor and acceptor capacities of the solvent and solute, the well recognized factors defining solubility of polar molecules and ions [42–45].

In contrast, more simple ASA-based implicit solvent methods focus on the first solvation shell effects that include short-range hydrophobic and hydrogen-bonding interactions with solvent. These methods calculate transfer energy as a product of ASA and solvation parameters for different types of atoms. Atomic solvation parameters, which describe transfer of specific atoms or groups from water or vapor to various solvents, such as octanol or cyclohexane, are usually determined by fitting to experimental transfer energies for a series of model compounds. Obtained solvation parameters are significantly different for nonpolar (hydrophobic) and polar (hydrogen-bonding) atoms [34, 46–48]. This approach has two deficiencies. First, the electrostatic energy is included as a part of the ASA-dependent contribution rather than a function of atomic charges and dipoles. Several models were developed to overcome this problem by combining the electrostatic energy with atomic solvation parameters [49–51]. Another problem is an oversimplified treatment of the lipid bilayer as a hydrophobic slab with membrane interfaces approximated by smoothing functions. Despite all simplifications, simulations with implicit solvent models were shown to reproduce membrane-binding modes of proteins and peptides with a reasonable precision [35, 38, 48, 52, 53].

At the other end of the spectrum of computational methods are approaches that use rigid-body optimization of proteins and simplified scoring functions for nonpolar amino acid residues [54, 55] or membrane depth-dependent whole-residue potentials derived from statistical analysis of membrane protein structures [56–60]. Some of these methods were implemented in publicly available web tools, TMDET [61] and Ez-3D [59] servers, which are fast and easy to use. Therefore, they can be applied for large-scale database scanning to identify integral membrane proteins and orient them in the lipid bilayer. Ez-3D server can also be used to evaluate orientation parameters and pseudo-energy landscapes of α-helical peptides in membrane. The calculated statistical pseudo-energies might correlate with actual membrane-binding energies of peptides; however, this was not verified using experimental data. Both TMDET and Ez methods were developed for integral membrane proteins and are not suitable for positioning of peripheral proteins weakly bound to membrane surface, although Ez-3D server demonstrates good predictions for peptides and some peripheral proteins that anchor to membranes through exposed amphiphilic α-helices. Besides, Ez-3D server occasionally has problems with TM β-barrels, which led to the development of Ezβ-methods for β-barrel proteins [56].

Motivated by the lack of a fast and reliable method for accurate prediction of both binding energies and the spatial arrangement of proteins in membranes, we have developed a new method for positioning of proteins in membranes (PPM) [48]. It provides rigid-body optimization in membranes of proteins with known 3D structures. As compared to other methods, the recently advanced PPM 2.0 version [62] implements two important new features. First, it uses a universal solvation model that describes transfer of solutes from water to an arbitrary fluid solvent with defined bulk polarity properties [63]. This solvation model operates not only with dielectric constant (ε) but also with Abraham's hydrogen-bonding acidity and basicity parameters (α and β [42]) of the solvent which were previously used in SMx solvation models [21, 64]. This model accounts for both long-range electrostatic and first solvation shell components of the transfer energy. Second, it represents the lipid bilayer as a fluid anisotropic solvent whose physico-chemical properties are defined by polarity profiles of parameters α, β, and ε along the membrane normal (Fig. 1). These profiles were calculated from experimental distributions of lipid segments in different lipid bilayers [65] instead of using an arbitrary system of hydrophobic slabs. Examination of non-monotonous changes of polarity parameters across the membrane reveals the existence of three distinct regions (Fig. 1): (1) the head group region at ~15–30 Å distance from the membrane center, where water concentration changes from 55.5 to 30 M; (2) the lipid core region at ~8 Å distance from the membrane center with cyclohexane-like properties ($\varepsilon \sim 2.5$) and low water concentration (<0.5 M); and (3) the "mid-polar" region at 8–15 Å distance from the membrane center with octanol-like properties ($\varepsilon \sim 11$) and intermediate water concentration (~1–3 M) [62]. The existence of the water-rich "mid-polar" region in membranes can explain the accumulation of amphiphilic residues (Tyr and Trp) near the membrane interface [66, 67] and the relatively low energy cost of insertion of polar and charged residues in this area, which is particularly important for the voltage-dependent movement of the voltage sensor, the Arg-rich S4-helix of the KvAP channel [62, 68].

The PPM 2.0 method was shown to accurately predict experimental transfer energies of small molecules from any solvent with defined bulk properties to water with root-mean-square errors (rmse) for neutral compounds in ions of 0.82 and 1.61 kcal/mol, respectively [63]. The application of the method for positioning of molecules in membranes was validated against experimental data

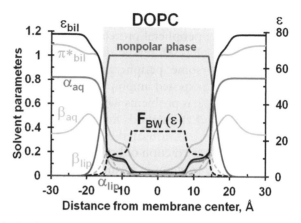

Fig. 1 Anisotropic solvent model of the lipid bilayer. Profiles of hydrogen bonding donor (α_1 and α_2) and acceptor (β_1 and β_2) capacities, solvatochromic dipolarity parameter (π^*), Block–Walker dielectric function, $F_{BW}(\varepsilon)$, and hydrocarbon profile were calculated for DOPC bilayer. "Mid-polar" region with octanol-like properties is colored *beige*

for 24 TM and 42 peripheral proteins, 30 peptides, and 23 small molecules [48, 53, 62]. Predicted orientation parameters and transfer-free energies were compared with experimental data for proteins and peptides, including membrane-binding energies, tilts in membranes, and sets of residues (or atoms) penetrating to the lipid bilayer that were assessed by spin-labeling, fluorescence, and chemical modification studies. Membrane-binding energies of peripheral proteins which do not associate specifically with lipids were calculated with rmse of 0.74 kcal/mol from the corresponding experimental values [62].

Here we present the general concept of our PPM 2.0 method, its verified applications, and the detailed protocol for using our PPM server which predicts the positions of peptides and proteins with known 3D structure in membrane and calculates their membrane-binding energies and orientation parameters.

2   Concept

2.1   Theory

The spatial positions of proteins in membranes are calculated using the PPM 2.0 method, which combines all-atom representation of a solute, a continuum anisotropic solvent approximation of the lipid bilayer, and a new universal solvation model that describes transfer from water to an arbitrary fluid solvent with defined bulk polarity properties [62]. This is a general physical method, which does not require a parameter adjustment for different classes of molecules. The PPM 2.0 method was made publicly available as a web server [4].

Transfer free energy of a protein from water to the lipid bilayer is calculated as the sum of three terms: (a) first solvation shell energy which depends on the solvent-accessible surface area and solvation parameter for each atom type; (b) long-range electrostatic contribution which depends on dipole moments of atomic groups in protein; and (c) long-range deionization energy of ionizable groups:

$$\Delta G_{\text{transf}}(d, D, \phi, \tau) = \sum_i [\sigma_i^{\text{wat} \to \text{bil}}(z_i) \text{ASA}_i + \eta^{\text{wat} \to \text{bil}}(z_i)\mu_i + \min\{E_i^{\text{ion}}, E_i^{\text{neutr}}\}]$$

where $\sigma_i^{\text{wat} \to \text{bil}}(z_i)$ is atomic solvation parameter that describes surface transfer energy (expressed in cal mol^{-1} Å^{-2}) of atom type i from water to position described by coordinates z_i along the bilayer normal, ASA_i is solvent-accessible surface area of atom i, $\eta^{\text{wat} \to \text{bil}}(z_i)$ is an energy cost of transferring the dipole of $1D$, μ_i is a group dipole moment, E_i^{ion} and E_i^{neutr} are energies of atom i in the ionized and neutral states, respectively, of an ionizable group.

The universal solvation model defines the first solvation shell contribution to transfer energy as a function of ASA and atomic solvation parameters (σ_i) which describe primarily the hydrophobic effect and hydrogen-bonding interactions. These parameters depend on atom types (via coefficients σ_i^0, e_i, a_i) and polarity properties of the solvent described by macroscopic dielectric constant (ε) and solvent hydrogen-bonding acidity (α) and basicity (β) parameters [42, 64].

$$\sigma_i^{\text{wat} \to \text{bil}}(z) = \sigma_i^0 - e_i(1/\varepsilon_{\text{wat}}) + a_i(\alpha_{\text{bil}}(z) - \alpha_{\text{wat}}) + b_i(\beta_{\text{bil}}(z) - \beta_{\text{wat}})$$

Coefficients σ_i^0, e_i, a_i, and b_i (and e_{dip}, below) for 26 types of atoms and ions were determined by linear regression of 1,269 experimental transfer energies from water to 19 solvents [63].

Long-range electrostatic contribution was described using a modified Born model for ionized groups [69] and Block–Walker dielectric function of solvent (F_{BW}) for dipoles [70]. The dipolar parameter η was found to be linearly dependent on the Block–Walker function:

$$\eta^{\text{wat} \to \text{bil}}(z) = e_{\text{dip}}(F_{\text{BW}}^{\text{bil}}(z) - F_{\text{BW}}^{\text{Wat}})$$

The method also accounts for the preferential solvation of protein groups by water and the hydrophobic mismatch for TM proteins [62].

The anisotropic properties of the lipid bilayer are described by transbilayer profiles of dielectric constant and hydrogen-bonding acidity and basicity parameters, $\varepsilon(z)$, $\alpha(z)$, and $\beta(z)$. We use polarity profiles of 1,2-dioleoyl-*sn*-glycero-3-phosphocholine (DOPC) bilayer derived from experimental distributions of quasi-molecular segments of lipids determined by neutron and X-ray scattering [65], and distribution of water across the DOPC bilayer determined in spin-labeling experiments [71] (Fig. 1).

The spatial location of a protein in the membrane coordinate system is obtained by rigid-body optimization of protein transfer

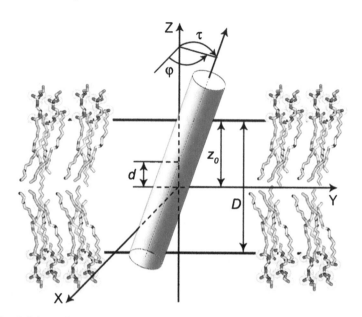

Fig. 2 Schematic representation of a transmembrane protein within membrane boundaries. Parameters used in rigid-body optimization: d, shift along the bilayer normal; D, hydrophobic thickness ($D=2z_0$); φ, rotation angle; τ, tilt angle

energy from water to the lipid bilayer. The position is optimized using a combination of grid scan and local energy minimization. The arrangement of a protein with respect to the membrane depends on its shift along the bilayer normal (d), rotational and tilt angles (φ and τ), and hydrophobic thickness of its membrane-spanning region for transmembrane proteins (D) (Fig. 2).

2.2 Applications   The verified applications of the PPM method include: (a) calculation of partition coefficients of solutes between any isotropic fluids [63]; (b) prediction of orientations in membranes of small molecules, peptides, and proteins with known structure [62], including newly solved crystal structures [72], (c) analysis of possible biological functions for proteins from Structural Genomics projects [73, 74]; (d) planning and interpretation of mutagenesis studies of membrane proteins [75]; and (e) modeling of partially unfolded membrane-bound protein states [76].

For example, PPM has been recently applied to study mechanism of transfer of α-tocopherol between cellular membranes by the α-tocopherol-transfer protein (αTTP) [75]. The method was able to quantitatively predict effects of mutations of solvent-exposed hydrophobic residues of αTTP involved in membrane association. Furthermore, PPM calculations in combination with FRET studies of genetically modified colicin E1 channel allowed us to propose a new umbrella model of the closed state of the colicin E1 channel domain [76]. PPM 2.0 was further applied for large-scale positioning in membranes of proteins and peptides from the entire PDB. Results of calculations were deposited in our OPM database [4] which currently contains more than 6,200 entries. The pre-oriented protein structures from the OPM database are being used by automated Membrane Builder [77] and Cell Microcosmos Membrane Editor [78] to produce starting structures of membrane–protein complexes for MD simulations, as well as for analysis and visualization of biomembranes.

Importantly, the all-atom approximation used in PPM can be easily converted into whole-residue hydrophobicity scales, which may be further applied for analysis and prediction of TM segments in proteins [62]. To obtain hydrophobicity scales for 18 amino acid residues (excluding Gly and Pro) for any location along the bilayer normal, we calculated transfer energy of each of these residues inserted into poly-Ala α-helix moving along the bilayer normal ("physical energy" profiles). We compared these "physical energy" profiles with "statistical energy" profiles derived from the relative abundance of these residues in lipid-exposed surface of 119 α-helical and 53 β-barrel membrane proteins. The "physical energy" profiles reproduced the shape of "statistical energy" profiles, including deep energy minima for Trp and Tyr in the "midpolar" region of the lipid bilayer.

We also found a good correlation between the "physical" and "statistical" scales for residues placed in the middle of the bilayer ($R^2 = 0.92$ for α-helical proteins and $R^2 = 0.91$ for β-barrels), as well as with the "biological" hydrophobicity scale determined from in vitro translocon-mediated insertion experiments [79] ($R^2 = 0.83$). However, "physical energies" for residues in the isolated poly-Ala α-helix were more than twofold larger than "statistical" and translocon-derived energies (the slopes of curves ΔG_{exp} vs. ΔG_{calc} were 0.46–0.48 and 0.39, respectively). These differences in the magnitude of the transfer free energies between PPM calculations, statistical analysis, and in vitro experiments can be explained by the two- to threefold larger ASA of these residues in a single α-helix used in PPM calculations as compared to the corresponding average ASA in polytopic proteins and in multi-helical Sec61 translocon machinery. Indeed, we estimated that ASA of lipid-exposed residues in multi-helical membrane proteins constitutes 30–60 % of their ASA in single α-helix and 20–35 % of ASA of corresponding side-chain analogues [62].

These observations suggest that two distinct whole-residue hydrophobicity scales should be used to predict TM segments in single-spanning (bitopic) and multi-spanning (polytopic) membrane proteins. Due to 2.5-fold decreased ASA of lipid-exposed residues in polytopic proteins, the energy penalties for membrane insertion of polar and charged residues used for polytopic proteins should be at least 2.5 times smaller than those used for bitopic proteins. However, these estimations are approximate, because the differences of ASA depend on residue type and structural context. Therefore, for the correct evaluation and optimization of the transfer-free energy of proteins of diverse structures, it is essential to employ all-atom approximation (as in PPM) rather than whole-residue hydrophobicity scales.

3 Methods

3.1 OPM Database

A large set of membrane-associated peptides and proteins from the PDB has been identified by PPM 2.0 and positioned with respect to the lipid bilayer using polarity profiles of the DOPC bilayer (Fig. 1). The protein coordinate files with predicted positions of membrane boundaries (at the level of lipid carbonyl groups) were collected in the OPM database (http://opm.phar.umich.edu/).

1. The OPM database includes only protein structures whose spatial positions in membranes can be computationally predicted, rather than a complete set of all membrane-interacting proteins from the PDB. Cα-atom models and some poor quality NMR models are not included, as well as peripheral proteins whose "membrane anchors" (amphiphilic helices or loops, lipidated residues, or specifically bound lipids) are disordered or missing in experimental structures.

2. To facilitate analysis of highly redundant PDB data, we selected one (rarely 2–3) PDB entry to be included into OPM as a representative of every protein. The representative structure is the most complete model of the protein with maximal number of domains or subunits and fewer disordered regions. Several representative structures were selected for one protein, if they significantly differ in conformation or quaternary structure. All other PDB entries of the same protein (e.g., corresponding to different protein–ligand complexes or mutants) were included as related entries. Each representative structure has an automatically generated page that displays classification, picture, orientation parameters (membrane penetration depths and tilt angle), internal and external links, downloadable coordinates, transmembrane segments, comments, and other information about the protein.

3. To allow phylogenetic analysis of membrane proteins, the OPM database provides classification of protein structures at four levels: type (TM, peripheral/monotopic proteins, and peptides), class (for example, α-helical polytopic, α-helical bitopic, β-barrel TM, and all-α, all-β, α+β, α/β peripheral proteins), superfamily (evolutionarily related proteins), and family (proteins with detectable sequence homology) [80]. OPM

provides its own manual classification based on SCOP [81], TCDB [80], Pfam [82], Uniprot [83], publications, and structural superposition of proteins. Families and superfamilies in OPM are linked to the corresponding Pfam families and clans and TCDB families. Multi-protein complexes are classified based on structure of their largest membrane-bound subunit or domain.

4. OPM currently includes almost all PDB models of membrane-interacting peptides (281 representatives, 633 entries, excluding some poor quality NMR models) and TM proteins (614 representatives, 1,478 entries, excluding only a few Cα-atom models), but a less complete set of monotopic/peripheral proteins (1,072 representative, 4,112 entries). The set of TM proteins is updated biweekly.

5. The OPM provides the following data for downloading: (a) coordinate files for individual protein structures with calculated membrane boundaries; (b) compressed coordinate files for several protein sets (α-helical polytopic proteins, α-helical bitopic proteins, β-barrel TM proteins, monotopic/peripheral proteins, and peptides); (c) lists of PDB codes that belong to all types, classes, superfamilies, and families; (d) list of all representative PDB entries; (e) list of TM subunits and secondary structure segments; and (f) MySQL tables of database content.

6. The OPM coordinate files of peptides and proteins (in PDB format) differ from the original PDB files in the following aspects: (a) oligomeric state of each protein is selected based on publications and generated by PISA [84], rather than taken directly from the PDB; (b) some missing side-chain atoms may be reconstructed; (c) some side-chain conformers are changed to optimize energy of protein–lipid interactions; (d) some flexible loops (usually in NMR-derived entries) are excluded; (e) proteins are arranged in the membrane coordinate system with Z axis corresponding to membrane normal, and the origin of coordinates corresponding to membrane center; and (f) locations of lipid carbonyl groups are marked by DUMMY atoms. Hence, we advise to use precalculated PDB entries provided in the OPM database rather than generate them using PPM server described below.

3.2 PPM Server

To predict spatial positions in membranes of newly determined protein structures or theoretical models that are not included in the OPM database, we designed the public PPM server (http://opm.phar.umich.edu/server.php). It can handle TM and peripheral proteins and peptides, including those with ligands and nonstandard amino acids. This is the only public web resource for predicting membrane-binding modes of peripheral proteins.

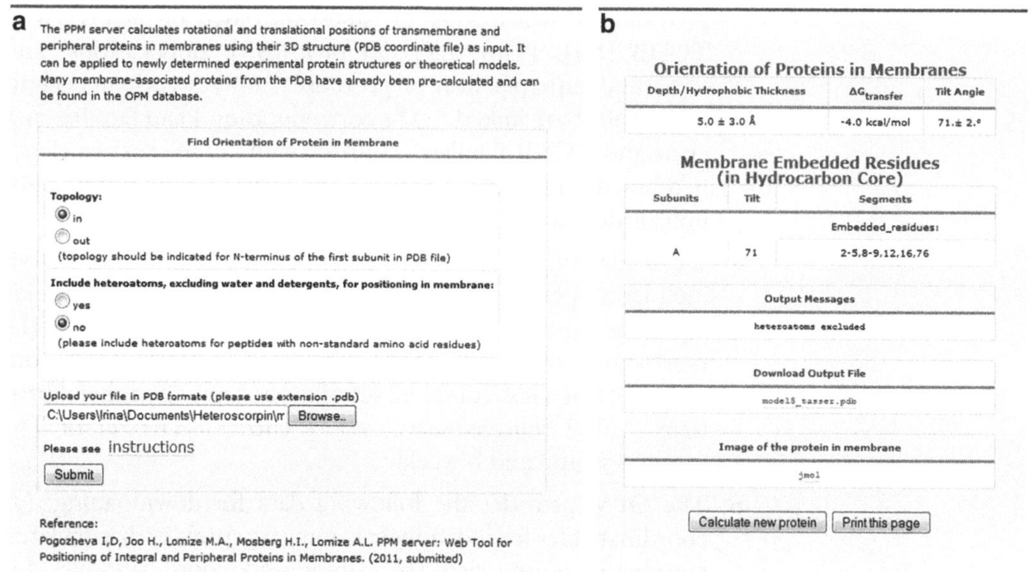

a

The PPM server calculates rotational and translational positions of transmembrane and peripheral proteins in membranes using their 3D structure (PDB coordinate file) as input. It can be applied to newly determined experimental protein structures or theoretical models. Many membrane-associated proteins from the PDB have already been pre-calculated and can be found in the OPM database.

Find Orientation of Protein in Membrane

Topology:

⊙ in

○ out

(topology should be indicated for N-terminus of the first subunit in PDB file)

Include heteroatoms, excluding water and detergents, for positioning in membrane:

○ yes

⊙ no

(please include heteroatoms for peptides with non-standard amino acid residues)

Upload your file in PDB formate (please use extension .pdb)

C:\Users\Irina\Documents\Heteroscorpin\m [Browse..]

Please see instructions

[Submit]

Reference:
Pogozheva I,D, Joo H., Lomize M.A., Mosberg H.I., Lomize A.L. PPM server: Web Tool for Positioning of Integral and Peripheral Proteins in Membranes. (2011, submitted)

b

Orientation of Proteins in Membranes

Depth/Hydrophobic Thickness	ΔG$_{transfer}$	Tilt Angle
5.0 ± 3.0 Å	-4.0 kcal/mol	71.± 2.°

Membrane Embedded Residues (in Hydrocarbon Core)

Subunits	Tilt	Segments
		Embedded_residues:
A	71	2-5,8-9,12,16,76

Output Messages

heteroatoms excluded

Download Output File

model5_tasser.pdb

Image of the protein in membrane

jmol

[Calculate new protein] [Print this page]

Fig. 3 Web interface for PPM server: input page (**a**) and output page (**b**)

Comparison with experimental data [4] demonstrated that PPM performs better than other similar servers [59, 61, 85].

3.2.1 Input and Data Preparation

1. Through the web interface of the PPM server (Fig. 3a), the user can upload the atomic coordinate file of a protein or a peptide that must be in the PDB format. Note that the name of the uploaded file must be a single word (no spaces, no special characters). It will be used to generate the name of the output file.

2. The user has additional options to: (a) specify topology of a protein ("in" or "out" for N-terminus of the first subunit in coordinate file) and (b) include ligands in calculation of transfer energy. The topology information is used to define the sign of Z coordinates for atoms (minus sign reserved for the "inner" side) and to assign atom type for DUMMY atoms that represent location of lipid carbonyl groups at the inner and outer leaflets (marked by "N" and "O" DUMMY atoms, respectively). The "ligand" option should be used to indicate the presence of nonstandard amino acids, lipids, cofactors, and other small molecules (labeled as HETATM) that should be in calculations. However, molecules of water and detergents are always automatically excluded.

3. To obtain more accurate results, the user is advised to: (a) reconstruct missing side-chain atoms that may interact with lipids; (b) remove His tags or other artificial chemical modifica-

tions; and (c) remove disordered loops with highly undefined spatial positions from NMR models. These preparations can be done using molecular modeling software (e.g., PyMol [86]).

4. The user is advised to provide the biologically relevant oligomeric state of a protein. The protein quaternary structures are precalculated by PISA (http://www.ebi.ac.uk/msd-srv/prot_int/) and can be downloaded from PDBe. However, PISA frequently generates several alternative assemblies that are not necessarily biologically relevant. Therefore, the correct biological units must be always verified through publications.

3.2.2 Output

1. Calculations of protein position in membranes may take from seconds to a few minutes, depending on the protein size and the workload on the server. Calculations of transmembrane proteins take longer than peripheral proteins (*see* **Note 1**).

2. The output (Fig. 3b) includes atomic coordinates of the protein with the predicted positions of lipid carbonyl groups indicated by DUMMY atoms. The output coordinate file (in PDB format) can be downloaded or viewed with JMol [87] (*see* **Note 2**).

3. The output window also displays a table with calculated orientation parameters: membrane penetration depth for peripheral proteins or hydrophobic thickness for TM proteins (Å), tilt angle (°), and water-to-membrane transfer energy (ΔG_{trans}, kcal/mol). Tilt angle is calculated between membrane normal (Z axis) and protein axis. The protein axis is defined as vector sum of TM secondary structure segment vectors (for TM proteins) or as the principal inertia axis (for peripheral proteins). The ± values for the depth and tilt angle show fluctuations of the corresponding parameters within 1 kcal/mol around the global minimum of transfer energy (*see* **Notes 3–5**).

4. The output also provides TM segments and a list of membrane-embedded residues (*see* **Note 6**).

4 Notes

1. The predicted orientations of peripheral membrane proteins can be verified using several criteria: (a) predicted orientations are consistent for different crystal structures of the same protein (some differences can be observed due to different conformations of side-chains and loops); (b) predicted membrane-binding modes are usually similar for different proteins from the same protein family or superfamily; (c) calculated membrane boundaries are expected to be spatially close

to identified binding sites for lipids or other hydrophobic ligands, lipidated residues, or predicted TM helices (for water-soluble domains of TM proteins).

2. The results of calculations are dependent on the quality of experimental structures. In particular, NMR models of peptides may have poorly defined loops or distorted secondary structures, which make calculated orientations highly unreliable and dependent on the selected NMR model or changes of side-chain conformations.

3. The values of transfer energy (ΔG_{transf}) calculated by PPM 2.0 are not especially informative for TM proteins, because insertion of these proteins is naturally assisted by the translocon machinery. The large values of ΔG_{transf} correlate with the large nonpolar surface of TM proteins exposed to the lipid bilayer. However, differences in ΔG_{transf} values between distinct conformations of the same TM protein are meaningful.

4. In contrast, the absolute ΔG_{transf} values for peripheral proteins correspond to their binding energies ($\Delta G_{binding}$) to the DOPC bilayer with a rmse of ~1 kcal/mol [62]. However, deviations can be larger in the following cases: (a) some membrane-anchoring elements of the protein are missing in the crystal structure because they are highly flexible or truncated; (b) the protein is anchored to membrane through specific binding to certain lipids ("lipid clams"); and (c) significant conformational changes (helix–coil transition or opening of a "closed" protein conformation) occur during membrane binding. These factors are not considered in calculation of ΔG_{transf}. The calculated values of ΔG_{transf} can be significantly smaller than experimental $\Delta G_{binding}$ in the first two cases, but larger in the last case [48, 62].

5. Results of calculations for peripheral proteins with transfer energies smaller than −5 kcal/mol must be interpreted with caution. Energies smaller than −3 kcal/mol indicate essentially no binding under physiological conditions due to the relatively low concentration of membranes in cell [7]. In such cases the calculated orientation may correspond to an intermediate weakly bound state which may lead to a stronger association upon conformational changes or ligand binding. In general, the calculations identify convex surfaces that are enriched in the solvent-exposed nonpolar and aromatic residues and may be involved in either protein–lipid or protein–protein interaction.

6. The PPM server produces diagnostic messages in a separate window. Any questions should be addressed to the developer (almz@umich.edu).

Acknowledgment

This work was supported by the Division of Biological Infrastructure of the National Science Foundation.

References

1. Israelachvili JN (1992) Intermolecular and surface forces. Academic, London
2. Leckband D, Israelachvili J (2001) Intermolecular forces in biology. Q Rev Biophys 34:105–267
3. Berman HM, Battistuz T, Bhat TN et al (2002) The protein data bank. Acta Crystallogr D: Biol Crystallogr 58:899–907
4. Lomize MA, Pogozheva ID, Joo H et al (2012) OPM database and PPM web server: resources for positioning of proteins in membranes. Nucleic Acids Res 40:D370–D376
5. Stansfeld PJ, Sansom MSP (2011) From coarse grained to atomistic: a serial multiscale approach to membrane protein simulations. J Chem Theory Comput 7:1157–1166
6. White SH, Wimley WC, Ladokhin AS et al (1998) Protein folding in membranes: determining energetics of peptide-bilayer interactions. Methods Enzymol 295:62–87
7. Wimley WC (2010) Energetics of peptide and protein binding to lipid membranes. Adv Exp Med Biol 677:14–23
8. Johansson ACV, Lindahl E (2009) The role of lipid composition for insertion and stabilization of amino acids in membranes. J Chem Phys 130:185101, http://dx.doi.org/10.1063/1.3129863
9. Ash WL, Zlomislic MR, Oloo EO et al (2004) Computer simulations of membrane proteins. Biochim Biophys Acta 1666:158–189
10. Mackerell AD (2004) Empirical force fields for biological macromolecules: overview and issues. J Comput Chem 25:1584–1604
11. Guvench O, MacKerell ADJ (2008) Comparison of protein force fields for molecular dynamics simulations. Methods Mol Biol 443:63–88
12. Bordner AJ (2012) Force fields for homology modeling. Methods Mol Biol 857:83–106
13. Domanski J, Stansfeld PJ, Sansom MSP et al (2010) Lipidbook: a public repository for force-field parameters used in membrane simulations. J Membr Biol 236:255–258
14. Raval A, Piana S, Eastwood MP et al (2012) Refinement of protein structure homology models via long, all-atom molecular dynamics simulations. Proteins. doi:10.1002/prot.24098
15. Koehl P, Levitt M (1999) A brighter future for protein structure prediction. Nat Struct Biol 6:108–111
16. Yesylevskyy SO, Schafer LV, Sengupta D et al (2010) Polarizable water model for the coarse-grained MARTINI force field. PLoS Comput Biol 6:e1000810. doi1000810.1001371/journal.pcbi.1000810.
17. Halgren TA, Damm W (2001) Polarizable force fields. Curr Opin Struct Biol 11:236–242
18. Lazaridis T, Archontis G, Karplus M (1995) Enthalpic contribution to protein stability: insights from atom-based calculations and statistical mechanics. Adv Protein Chem 47:231–306
19. Kollman P (1993) Free-energy calculations: applications to chemical and biochemical phenomena. Chem Rev 93:2395–2417
20. Cramer CJ, Truhlar DG (1999) Implicit solvation models: equilibria, structure, spectra, and dynamics. Chem Rev 99:2161–2200
21. Cramer CJ, Truhlar DG (2008) A universal approach to solvation modeling. Acc Chem Res 41:760–768
22. Woolf TB, Roux B (1994) Molecular-dynamics simulation of the gramicidin channel in a phospholipid-bilayer. Proc Natl Acad Sci USA 91:11631–11635
23. Knecht V, Grubmuller H (2003) Mechanical coupling via the membrane fusion SNARE protein syntaxin 1A: a molecular dynamics study. Biophys J 84:1527–1547
24. Tozzini V (2005) Coarse-grained models for proteins. Curr Opin Struct Biol 15:144–150
25. Marrink SJ, Risselada HJ, Yefimov S et al (2007) The MARTINI force field: coarse grained model for biomolecular simulations. J Phys Chem B 111:7812–7824
26. Stansfeld PJ, Sansom MSP (2011) Molecular simulation approaches to membrane proteins. Structure 19:1562–1572
27. Rouse S, Carpenter T, Sansom MSP (2010) Coarse-grained molecular dynamics simulations of membrane proteins. In: Sansom MSP, Biggin PC (eds) Molecular simulations and biomembranes: from biophysics to function, vol 20, 1st edn. RSC Biomolecular Sciences, Cambridge, UK, pp 56–75

28. Hall BA, Chetwynd AP, Sansom MSP (2011) Exploring peptide-membrane interactions with coarse-grained MD simulations. Biophys J 100:1940–1948

29. Chetwynd A, Wee CL, Hall BA et al (2010) The energetics of transmembrane helix insertion into a lipid bilayer. Biophys J 99:2534–2540

30. Chetwynd AP, Scott KA, Mokrab Y et al (2008) CGDB: a database of membrane protein/lipid interactions by coarse-grained molecular dynamics simulations. Mol Membr Biol 25:662–669

31. Lindahl E, Sansom MSP (2008) Membrane proteins: molecular dynamics simulations. Curr Opin Struct Biol 18:425–431

32. Sherwood P, Brooks BR, Sansom MSP (2008) Multiscale methods for macromolecular simulations. Curr Opin Struct Biol 18:630–640

33. Grossfield A (2008) Implicit modeling of membranes. In: Feller SE (ed) Computational modeling of membrane bilayers, vol 60, 1st edn, Current topics in membranes. Academic, London, pp 131–157

34. Bordner AJ, Zorman B, Abagyan R (2011) Efficient molecular mechanics simulations of the folding, orientation, and assembly of peptides in lipid bilayers using an implicit atomic solvation model. J Comput Aided Mol Des 25:895–911

35. Ulmschneider MB, Ulmschneider JP, Sansom MSP et al (2007) A generalized born implicit-membrane representation compared to experimental insertion free energies. Biophys J 92:2338–2349

36. Tanizaki S, Feig M (2006) Molecular dynamics simulations of large integral membrane proteins with an implicit membrane model. J Phys Chem B 110:548–556

37. Feig M, Brooks CL (2004) Recent advances in the development and application of implicit solvent models in biomolecule simulations. Curr Opin Struct Biol 14:217–224

38. Im W, Feig M, Brooks CL (2003) An implicit membrane generalized born theory for the study of structure, stability, and interactions of membrane proteins. Biophys J 85:2900–2918

39. Roux B, Simonson T (1999) Implicit solvent models. Biophys Chem 78:1–20

40. Sitkoff D, Sharp KA, Honig B (1994) Accurate calculation of hydration free-energies using macroscopic solvent models. J Phys Chem 98:1978–1988

41. Chen J, Brooks CL (2008) Implicit modeling of nonpolar solvation for simulating protein folding and conformational transitions. Phys Chem Chem Phys 10:471–481

42. Abraham MH (1993) Scales of solute hydrogen-bonding—their construction and application to physicochemical and biochemical processes. Chem Soc Rev 22:73–83

43. Marcus Y (1998) Some thermodynamic aspects of ion transfer. Electrochim Acta 44:91–98

44. Abraham MH, Zhao YH (2004) Determination of solvation descriptors for ionic species: hydrogen bond acidity and basicity. J Org Chem 69:4677–4685

45. Reichardt C (2007) Solvents and solvent effects: an introduction. Org Process Res Dev 11:105–113

46. Ducarme P, Rahman M, Brasseur R (1998) IMPALA: a simple restraint field to simulate the biological membrane in molecular structure studies. Proteins 30:357–371

47. Efremov RG, Nolde DE, Vergoten G et al (1999) A solvent model for simulations of peptides in bilayers. I. Membrane-promoting alpha-helix formation. Biophys J 76:2448–2459

48. Lomize AL, Pogozheva ID, Lomize MA et al (2006) Positioning of proteins in membranes: a computational approach. Protein Sci 15:1318–1333

49. Bordner AJ, Cavasotto CN, Abagyan RA (2002) Accurate transferable model for water, n-octanol, and n-hexadecane solvation free energies. J Phys Chem B 106:11009–11015

50. Lazaridis T (2003) Effective energy function for proteins in lipid membranes. Proteins 52:176–192

51. Lazaridis T (2005) Implicit solvent simulations of peptide interactions with anionic lipid membranes. Proteins 58:518–527

52. Efremov RG, Nolde DE, Konshina AG et al (2004) Peptides and proteins in membranes: what can we learn via computer simulations? Curr Med Chem 11:2421–2442

53. Lomize AL, Pogozheva ID, Lomize MA et al (2007) The role of hydrophobic interactions in positioning of peripheral proteins in membranes. BMC Struct Biol 7:44. doi:10.1186/1472-6807-1187-1144

54. Tusnady GE, Dosztanyi Z, Simon I (2005) PDB_TM: selection and membrane localization of transmembrane proteins in the protein data bank. Nucleic Acids Res 33:D275–D278

55. Tusnady GE, Dosztanyi Z, Simon I (2004) Transmembrane proteins in the protein data bank: identification and classification. Bioinformatics 20:2964–2972

56. Hsieh D, Davis A, Nanda V (2012) A knowledge-based potential highlights unique features of membrane α-helical and β-barrel

protein insertion and folding. Protein Sci 21:50–62

57. Ulmschneider MB, Sansom MSP, Di Nola A (2005) Properties of integral membrane protein structures: derivation of an implicit membrane potential. Proteins 59:252–265

58. Senes A, Chadi DC, Law PB et al (2007) E-z, a depth-dependent potential for assessing the energies of insertion of amino acid side-chains into membranes: derivation and applications to determining the orientation of transmembrane and interfacial helices. J Mol Biol 366:436–448

59. Schramm CA, Hannigan BT, Donald JE et al (2012) Knowledge-based potential for positioning membrane-associated structures and assessing residue-specific energetic contributions. Structure 20:924–935

60. Ulmschneider MB, Sansom MSP, Di Nola A (2006) Evaluating tilt angles of membrane-associated helices: comparison of computational and NMR techniques. Biophys J 90: 1650–1660

61. Tusnady GE, Dosztanyi Z, Simon I (2005) TMDET: web server for detecting transmembrane regions of proteins by using their 3D coordinates. Bioinformatics 21:1276–1277

62. Lomize AL, Pogozheva ID, Mosberg HI (2011) Anisotropic solvent model of the lipid bilayer. 2. Energetics of insertion of small molecules, peptides, and proteins in membranes. J Chem Inf Model 51:930–946

63. Lomize AL, Pogozheva ID, Mosberg HI (2011) Anisotropic solvent model of the lipid bilayer. 1. Parameterization of long-range electrostatics and first solvation shell effects. J Chem Inf Model 51:918–929

64. Li JB, Zhu TH, Hawkins GD, Winget P et al (1999) Extension of the platform of applicability of the SM5.42R Universal solvation model. Theor Chem Acc 103:9–63

65. Kucerka N, Nagle JF, Sachs JN et al (2008) Lipid bilayer structure determined by the simultaneous analysis of neutron and x-ray scattering data. Biophys J 95:2356–2367

66. Yau WM, Wimley WC, Gawrisch K et al (1998) The preference of tryptophan for membrane interfaces. Biochemistry 37:14713–14718

67. Killian JA, von Heijne G (2000) How proteins adapt to a membrane-water interface. Trends Biochem Sci 25:429–434

68. Freites JA, Tobias DJ, von Heijne G, White SH (2005) Interface connections of a transmembrane voltage sensor. Proc Natl Acad Sci USA 102:15059–15064

69. Abe T (1986) A modification of the born equation. J Phys Chem 90:713–715

70. Block H, Walker SM (1973) Modification of Onsager theory for a dielectric. Chem Phys Lett 19:363–364

71. Marsh D (2002) Membrane water-penetration profiles from spin labels. Eur Biophys J 31:559–562

72. Rufer AC, Lomize A, Benz J et al (2007) Carnitine palmitoyltransferase 2: analysis of membrane association and complex structure with a substrate analog. FEBS Lett 581: 3247–3252

73. Chiu HJ, Bakolitsa C, Skerra A, Lomize A et al (2010) Structure of the first representative of pfam family PF09410 (DUF2006) reveals a structural signature of the calycin superfamily that suggests a role in lipid metabolism. Acta Crystallogr Sect F Struct Biol Cryst Commun 66:1153–1159

74. Kumar A, Lomize A, Jin KK et al (2010) Open and closed conformations of two SpoIIAA-like proteins (YP_749275.1 and YP_001095227.1) provide insights into membrane association and ligand binding. Acta Crystallogr Sect F Struct Biol Cryst Commun 66:1245–1253

75. Zhang WX, Thakur V, Lomize A et al (2011) The contribution of surface residues to membrane binding and ligand transfer by the alpha-tocopherol transfer protein (alpha-TTP). J Mol Biol 405:972–988

76. Ho D, Lugo MR, Lomize AL et al (2011) Membrane topology of the colicin E1 channel using genetically encoded fluorescence. Biochemistry 50:4830–4842

77. Jo S, Kim T, Im W (2007) Automated builder and database of protein/membrane complexes for molecular dynamics simulations. PLoS One 2:e880. doi:e880 810.1371/journal.pone.0000880

78. Sommer B, Dingersen T, Gamroth C et al (2011) CELLmicrocosmos 2.2 membrane editor: a modular interactive shape-based software approach to solve heterogeneous membrane packing problems. J Chem Inf Model 51:1165–1182

79. Hessa T, Kim H, Bihlmaier K et al (2005) Recognition of transmembrane helices by the endoplasmic reticulum translocon. Nature 433:377–381

80. Saier MH, Yen MR, Noto K et al (2009) The Transporter classification database: recent advances. Nucleic Acids Res 37:D274–D278

81. Andreeva A, Howorth D, Chandonia JM et al (2008) Data growth and its impact on the SCOP database: new developments. Nucleic Acids Res 36:D419–D425

82. Bateman A, Birney E, Cerruti L et al (2002) The pfam protein families database. Nucleic Acids Res 30:276–280

83. Apweiler R, Martin MJ, O'Donovan C et al (2010) The Universal protein resource (UniProt) in 2010. Nucleic Acids Res 38:D142–D148

84. Krissinel E, Henrick K (2007) Inference of macromolecular assemblies from crystalline state. J Mol Biol 372:774–797

85. Cheema J, Basu G (2011) MAPS: an interactive web server for membrane annotation of transmembrane protein structures. Indian J Biochem Biophys 48:106–110

86. DeLano WL (2003) The PyMOL molecular graphics system. DeLano Scientific LLC, San Carlos, CA, http://www.pymol.org/

87. Bond PJ, Sansom MSP (2006) Insertion and assembly of membrane proteins via simulation. J Am Chem Soc 128:2697–2704

Part III

NMR Methods

Chapter 8

Membrane Protein Structure Determination: Back to the Membrane

Yong Yao, Yi Ding, Ye Tian, Stanley J. Opella, and Francesca M. Marassi

Abstract

NMR spectroscopy enables the structures of membrane proteins to be determined in the native-like environment of the phospholipid bilayer membrane. This chapter outlines the methods for membrane protein structural studies using solid-state NMR spectroscopy with samples of membrane proteins incorporated in proteoliposomes or planar lipid bilayers. The methods for protein expression and purification, sample preparation, and NMR experiments are described and illustrated with examples from OmpX and Ail, two bacterial outer membrane proteins that function in bacterial virulence.

Key words Membrane protein, NMR, Lipid, Bilayer, Membrane, Protein, Expression, Structure, Barrel

1 Introduction

Membrane proteins mediate all interactions of a cell or organism with the outside world and, as such, are responsible for the basic human experiences (taste, smell, touch, sight, thought, etc.) that constitute life. They are encoded by ~30 % of all known pro- or eukaryotic genes and perform essential biological functions that include cellular transport, signaling, and programmed cell death. Dysfunctions of human membrane proteins are linked with devastating diseases, and the membrane proteins encoded by viruses and bacteria are major players in infection, virulence, or antibiotic resistance. It is, therefore, not surprising that membrane proteins are the principal targets of all drugs on the market today and that understanding their biological functions is a major goal of biomedical research in academic, medical, biotech, and pharmaceutical settings.

Despite their importance, very little structural data exist for them compared to the wealth of information available for their soluble counterparts. This reflects the special amphiphilic properties of membrane proteins and their surrounding membrane environment, which complicate biophysical studies. Membrane proteins differ fundamentally from water-soluble proteins. While the latter

Giovanna Ghirlanda and Alessandro Senes (eds.), *Membrane Proteins: Folding, Association, and Design*,
Methods in Molecular Biology, vol. 1063, DOI 10.1007/978-1-62703-583-5_8, © Springer Science+Business Media, LLC 2013

exist in an isotropic aqueous environment, the lipid bilayer membrane is anisotropic and heterogeneous, with large gradients in fluidity, water concentration, and dielectric constants from the bilayer core to the water-lipid interface [1–4]. These features lead to phenomena (e.g., stronger hydrogen bonds, membrane thinning, hydrophobic mismatch, curvature frustration, charge polarization, lateral force gradients) that significantly influence membrane protein structure, dynamics, and function and argue very strongly in favor of determining the structures of membrane proteins in lipid bilayers at or near physiological conditions of temperature, pH, and hydration [reviewed in ref. 5–7]. As noted by Cross [7], this is in line with Anfinsen's hypothesis, which states that a protein conformation "is determined by the totality of inter-atomic interactions and hence by the amino acid sequence in a given environment" [8]. For membrane proteins, the "given environment" of the lipid bilayer is essential for preserving native structure and function.

Membrane protein structure determination by X-ray diffraction and solution NMR requires proteins dissolved in detergents because lipid bilayers are incompatible with crystallization and solubilization. For some membrane proteins, lipid nanodiscs can be useful membrane mimics for solution NMR [9–11], but typically they yield broader lines and significant sample polydispersity compared to micelles or bicelles. In contrast, solid-state NMR is compatible with structure determination of membrane proteins in membranes, under physiological conditions, and recent developments in sample preparation, recombinant bacterial expression, pulse sequences for high-resolution NMR spectroscopy, and computational methods have enabled a number of membrane protein structures to be determined in lipid bilayer membranes (Fig. 1; [12, 13]).

In this chapter, these methods are illustrated with examples from two homologous bacterial outer membrane proteins, OmpX (outer membrane protein X) from *E. coli* and Ail (attachment invasion locus) from *Yersinia pestis*, an extremely pathogenic organism

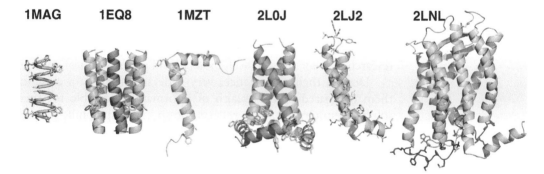

Fig. 1 Solid-state NMR structures of membrane proteins in lipid bilayers. PDB codes correspond to (1MAG) gramicidin [94], (1EQ8) M2 pore-forming domain of acetylcholine receptor [95], (1MZT) membrane-bound bacteriophage fd coat protein [25], (2L0J) channel-forming domain of influenza M2 [34], (2LJ2) mercury transporter MerF [33], and (2LNL) human chemokine receptor CXCR1 [54]

with a long history of precipitating massive human pandemics of plague [14]. Although the specific function of OmpX is not known, Ail is a *Y. pestis* virulence factor essential for evading the human host's immune system by mediating the adhesion of *Y. pestis* to human host cells and providing resistance to human innate immunity [15, 16]. Both OmpX and Ail belong to a family (pfam PF06316) of outer membrane proteins that share amino acid sequence homology in the membrane-spanning segments but vary widely in the sequences of the four extracellular loops, which are critical for the function of Ail. Both proteins adopt a transmembrane 8-stranded β-barrel structure [17–19]. However, while all extracellular loops of OmpX are fully structured, key functional loops of Ail are not visible in the crystal structure, indicating that the structure of Ail determined within the phospholipid bilayer membrane will be needed to understand the molecular basis of its biological function.

2 Materials

Specialized materials used for the experiments described in this chapter are listed in Table 1. They include *E. coli* cells for recombinant expression, lipids for protein reconstitution, and isotopically labeled salts, sugars, and amino acids used to produce ^{15}N- and ^{13}C-labeled proteins for NMR studies.

Table 1
Specialized materials and computer programs

Material	Source
Reagents	
E. coli BL21(DE3) cells	Novagen (www.emdmillipore.com)
IPTG (isopropyl 1-thio-β-D-galactopyranoside)	Sigma (www. sigmaaldrich.com)
DHPC (dihexanoyl-phosphocholine)	Avanti Polar Lipids (www.avantilipids.com)
6-O-PC (di-O-hexyl-phosphocholine)	Avanti Polar Lipids (www.avantilipids.com)
DMPC (di-myristoyl-phosphocholine)	Avanti Polar Lipids (www.avantilipids.com)
14-O-PC (di-O-tetradecyl-phosphocholine)	Avanti Polar Lipids (www.avantilipids.com)
DMPG (di-myristoyl-phosphoglycerol)	Avanti Polar Lipids (www.avantilipids.com)
SDS (sodium-dodecyl-sulfate)	Sigma (www. sigmaaldrich.com)
(^{15}NH$_4$)$_2$SO$_4$	Cambridge Isotopes laboratories (www.isotope.com)
^{13}C-glucose	Cambridge Isotopes laboratories (www.isotope.com)
^{15}N-Val	Cambridge Isotopes laboratories (www.isotope.com)
^{15}N-Phe	Cambridge Isotopes laboratories (www.isotope.com)
YbCl$_3$	Sigma Aldrich (www.sigma.com)
HiTrap SP HP 5 mL column	GE Healthcare Life Sciences (www.gelifesciences.com)
Sephacryl S-200 HR HiPrep 16/60 column	GE Healthcare Life Sciences (www.gelifesciences.com)
Computer programs	
NMRPipe	(spin.niddk.nih.gov/bax/software)
Sparky	(www.cgl.ucsf.edu/home/sparky/)
XPLOR-NIH/AssignFit	(nmr.cit.nih.gov)
ROSETTA	(www.rosettacommons.org)

3   Methods

3.1   Protein Expression and Purification

Cloning. Expression, purification, and refolding of OmpX and Ail were performed as described previously [17, 20]. The genes encoding mature Ail and OmpX (without the signal sequence) were cloned between the NdeI and XhoI restriction sites of the pET-30b. For both OmpX and Ail, deletion of the signal sequence directs protein expression into inclusion bodies.

Protein Expression. The Ail- and OmpX-encoding plasmids were transformed in *E. coli* BL21 (DE3) cells. Positive clones were grown in 5 mL of LB medium at 37 °C for 8 h, then 100 μL of this culture was used to inoculate 50 mL of M9 minimal medium for overnight growth at 37 °C. The next morning, 50 mL of overnight culture was transferred to 1 L of fresh M9 medium, and the cells were grown at 37 °C, to a density of $OD_{600} = 0.6$, before induction with 1 mM IPTG for 3–6 h. Cells were harvested by centrifugation (6,500 rpm, 15 min, 4 °C) and stored at –80 °C overnight. For ^{15}N and ^{13}C isotopic labeling, the M9 growth medium was prepared with $(^{15}NH_4)_2SO_4$ and ^{13}C-glucose (Cambridge Isotope Laboratories) as the sole nitrogen and carbon sources.

Inclusion Bodies Isolation. Ail expression and purification were monitored by SDS-PAGE (Fig. 2). Cells from a 1 L culture were suspended in 30 mL of buffer A (Table 2) and lysed by two passes through a French press. The soluble cell fraction was removed by centrifugation (19,000 rpm, 30 min, 4 °C) and the remaining pellet, which is enriched in inclusion bodies, was suspended in 30 mL buffer A_T and gently mixed for 1 h at 37 °C. The soluble fraction was removed by centrifugation (19,000 rpm, 30 min, 4 °C), then the resulting pellet was washed with 30 mL of water to remove residual detergent and again isolated by centrifugation (19,000 rpm, 30 min, 4 °C). The isolated inclusion bodies are white in appearance

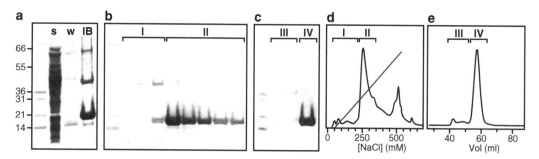

Fig. 2 Expression and purification of Ail. (**a–c**) SDS-PAGE and (**d, e**) chromatograms showing Ail purification steps. (**a**) Inclusion bodies isolation; the supernatant (s) and wash (w) fractions are free of Ail, while inclusion bodies (IB) are enriched in Ail. (**b, d**) Ion exchange chromatography yields a fraction free of Ail (I) and another containing Ail (II). (**c, e**) Size exclusion chromatography yields a fraction free of Ail (III) and a fraction containing purified Ail (IV)

Table 2
Buffers used for protein purification and sample preparation

Buffer A	20 mM Tris-HCl, pH 8.0
Buffer A$_T$	20 mM Tris-HCl, pH 8.0, 2 % Triton-X
Buffer B	20 mM Na acetate, pH 5.0, 8 M urea
Buffer B$_s$	20 mM Na acetate, pH 5.0, 8 M urea, 150 mM NaCl
Buffer C	20 mM Tris-HCl, pH 9.5, 150 mM NaCl
Buffer E	20 mM Tris-HCl, pH 8.0, 8 M urea
Buffer F	20 mM glycine, pH 10.2, 200 mM KCl

and contain very pure protein. Finally, the inclusion bodies pellet was dissolved in 30 mL of buffer B by gently mixing for 1 h at 37 °C, and the remaining insoluble fraction was removed by centrifugation (19,000 rpm, 30 min, 4 °C) and discarded (Fig. 2a).

Protein Purification. The purification strategy depends on each protein's properties. The isoelectric points of OmpX (pI = 5.0) and Ail (pI = 7.8) dictated the use of anion and cation exchange chromatography, respectively. For example, Ail was purified by cation exchange chromatography (HiTrap SP HP 5 mL column, GE Healthcare) in buffer B with a NaCl gradient (Fig. 2b, d), followed by size exclusion chromatography (Sephacryl S-200 HR HiPrep 16/60 column, GE Healthcare) in buffer B$_s$ (Fig. 2c, e). Purified protein was concentrated by dialysis (10 kDa cutoff) against water, lyophilized, and stored at –20 °C. Typically, 25–30 mg of purified protein is obtained from a 2 L culture in ^{15}N, ^{13}C isotopically labeled M9 medium.

3.2 Reconstitution in Phospholipid Bilayers

Pure Ail or OmpX (8 mg of lyophilized powder) was dissolved in 1 mL of 100 mM SDS in water, and added dropwise, at 40 °C, to a suspension of small unilamellar vesicles prepared by probe sonication with 50 mg of DMPC (Avanti Polar Lipids), or its ether-linked analog 14-O-PC, in 20 mL of buffer C. Although the chemical structure of these lipids is very similar, the ether link of 14-O-PC prevents lipid hydrolysis and is better suited for long-term sample stability. The protein/lipid mixture was incubated at 30 °C for 24 h and refolding was monitored by SDS-PAGE (Fig. 3). After complete refolding, SDS was removed by dialysis (10 kDa cutoff) against two 4 L changes of buffer C, followed by four 4 L changes of buffer C supplemented with 30 mM KCl. The resulting proteoliposomes were dialyzed against 4 L of buffer A and then concentrated by ultracentrifugation (41,000 rpm, 4 h, 4 °C). To prepare magnetically aligned planar bilayer samples, 10 mg of DHPC, or its ether-linked analog 6-O-PC, was dissolved in 50 μL of water,

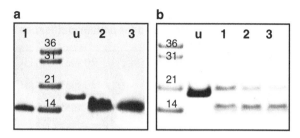

Fig. 3 Refolding of Ail and OmpX in lipids monitored by SDS-PAGE. Unfolded proteins (*lane u*) migrate at higher apparent molecular weights (~21 kDa) than folded proteins (~14 kDa). (**a**) Refolding of Ail in DMPC (*lane 1*), DHPC (*lane 2*), or DHPC/DMPC (*lane 3*). (**b**) Refolding of OmpX in DMPC monitored at 0 h (*lane 1*), 2 h (*lane 2*), and 12 h (*lane 3*) after addition from SDS solution. All samples were loaded without boiling

added to the proteoliposomes and thoroughly mixed by repeated freezing and thawing, as described previously [21, 22]. Alignment of the lipid bilayer normal parallel to the magnetic field was induced by adding 5 μL of 100 mM YbCl$_3$ directly to the NMR tube precooled at 4 °C, mixing thoroughly, and resealing the tube.

3.3 Solid-State NMR Studies in Lipid Bilayer Membranes

Approach for NMR Structural Studies. Modern NMR methods for protein structure determination increasingly rely on orientation restraints derived from dipolar coupling (DC) and chemical shift anisotropy (CSA) measurements, and on dihedral angle (ϕ, ψ) restraints derived from isotropic chemical shift (CS) analysis [23–28]. For both solution NMR and solid-state NMR, these restraints can be used to guide both the generation of structural models and structure refinement and are especially powerful when coupled with molecular fragment replacement (MFR) and de novo structure prediction using programs such as ROSETTA [29, 30]. By shifting the burden away from time-consuming measurements of multiple long-range distances between side chain sites, these approaches significantly facilitate protein structure determination and yield reliable three-dimensional structures, with very few or no distance restraints, for soluble proteins in water [27, 31], membrane proteins in micelles [32], and membrane proteins in lipid bilayers [25, 33, 34].

By combining features of magic-angle spinning (MAS) [35–37] and oriented-sample (OS) [38–40] solid-state NMR approaches, it is possible to resolve and assign multiple peaks through the use of $^{15}N/^{13}C$-labeled samples and to measure DC, CSA, and isotropic CS to obtain precise orientation and dihedral angle restraints for structure determination [33, 41]. OS solid-state NMR uses samples that are uniaxially aligned relative to the magnetic field (e.g., planar lipid bilayers) to yield orientation-dependent single-line resonances. MAS solid-state NMR uses nonaligned samples (e.g., proteoliposomes) and yields single-line spectra due to averaging of the spin interactions to their isotropic values.

In both cases, the uniaxial order inherent to membrane proteins undergoing rotational diffusion around the lipid bilayer normal [42–46] provides the foundation for a powerful approach to structure determination based on orientation restraints [41, 47–49]. The orientation-dependent DC and CSA signals correlate directly with molecular structure and enable both protein structure and global orientation (i.e., supramolecular structure) to be determined in the membrane [50–52]. Their frequencies can be read directly from the single-line resonances of OS solid-state NMR spectra, or MAS can be used to recouple and measure rotationally averaged powder patterns: since the frequency measured from the edge of a rotationally averaged powder pattern is equivalent to that measured from OS NMR spectra, the same analytical methods developed for data analysis and structure determination are applicable.

The DC and CSA interactions are well characterized for $^1H/^{15}N/^{13}C$-labeled protein sites [39, 40]. Their interpretation in terms of orientation restraints is greatly facilitated by the fact that solid-state NMR spectra display them at full, or near full, magnitudes, enabling a priori knowledge of the order tensor. Thus, a set of CSA and DC frequencies can provide sufficient restraints for high-precision structure determination, requiring few or no distances (e.g., [25, 33, 34, 53, 54]).

The spectra of uniaxially ordered samples exhibit distinctive wheel-like patterns [50–52] that reflect protein structure and orientation (Fig. 4). These patterns are observed for both α-helices [25, 50, 51, 55] and β-strands [21, 22, 52]. They stem from the direct relationship that exists between orientation-dependent solid-state NMR data and molecular structure and are useful both for guiding resonance assignment performed with traditional spectroscopic approaches (e.g., [56]) and for obtaining resonance assignments through methods that we have developed for simultaneous assignment and structure refinement (SASR) [25]. They also help reduce or eliminate the degeneracy of orientation solutions associated with DC and CSA measurements [25, 57], allowing structures to be built by linking consecutive peptide planes or fragments through their common CA atom.

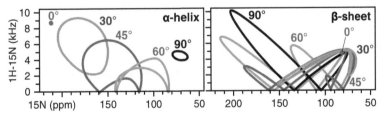

Fig. 4 Solid-state NMR *PISA wheels* observed in the OS solid-state NMR of α-helical and β-barrel membrane proteins. Theoretical *PISA wheels* are shown for ideal α-helices or β-strands with different tilts (0°, 30°, 45°, 60°, 90°) relative to the phospholipid bilayer membrane normal

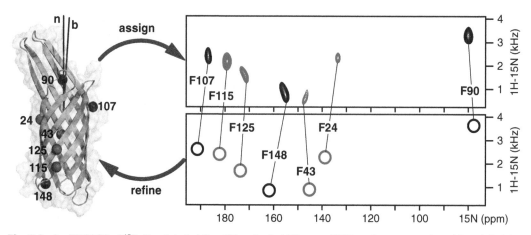

Fig. 5 AssignFit SASR of ¹⁵N-Phe-labeled OmpX in oriented bilayers. NMR peaks were assigned by minimizing the difference between experimental (*top*) and back-calculated (*bottom*) spectra after refinement of the OmpX crystal structure [17] with the assigned orientation restraints. The data provide the orientation of OmpX in the membrane (b: barrel axis; n: bilayer normal). Peaks in *red* persist in D_2O

Simultaneous Assignment and Structure Refinement. The direct relationship between NMR data and structure facilitates a method for SASR, based on minimizing the difference between the experimentally observed spectral frequencies and the frequencies back-calculated from a structural model [22, 25]. The SASR approach relieves the burden of having to obtain near complete resonance assignments prior to structure determination: resonance assignments are obtained as a side product of fitting a structural model to the NMR data, but is not a prerequisite for structure determination. To automate the SASR process, we have developed AssignFit, a Python-based program that is available as part of XPLOR-NIH release 2.29 [58]. AssignFit generates all assignment permutations and calculates the corresponding molecular alignment, the atomic coordinates reoriented in the alignment frame, and the associated set of NMR frequencies, which are then compared with the experimental data for best fit. For example, using AssignFit, the seven Phe peaks in the separated local field (SLF) spectrum of OmpX could be assigned easily and quickly [22, 58] (Fig. 5).

Solid-State NMR Experiments. For OmpX and Ail, MAS and OS solid-state NMR experiments were performed, on 500 or 700 MHz Bruker Avance spectrometers equipped with a Bruker low-E ¹H/¹³C/¹⁵N triple-resonance 3.2 mm MAS solid-state NMR probe (MAS experiments) or a static low-E ¹H/¹⁵N double-resonance solid-state NMR probe (OS experiments). During MAS experiments, the sample temperature was maintained at 25 ± 2 °C and the spinning rate was controlled to 11.11 ± 0.002 kHz using a Bruker

MAS controller. During OS experiments the sample temperature was maintained at 43 ± 2 °C. Chemical shifts were externally referenced to DSS, by setting the adamantane methylene carbons to a ^{13}C chemical shift frequency of 40.48 ppm, or to liquid ammonia, by setting the ammonium sulfate nitrogen ^{15}N chemical shift frequency to 26.8 ppm [59, 60]. The NMR data were processed and analyzed using NMRPipe [61] and Sparky [62].

Solid-State NMR Studies of Ail and OmpX in Proteoliposomes. MAS solid-state NMR studies benefit from the rapid progress made by numerous laboratories around the world [35–37, 63–68]. Resonances in these spectra can be assigned using NCACX and NCOCA experiments [69–74], complemented by ^{13}C-^{13}C [75, 76] and ^{13}C-^{15}N correlation experiments [77]. Two-dimensional ^{13}C-^{13}C correlation MAS spectra of Ail and OmpX in proteoliposomes (Fig. 6a, b) show several resolved peaks, and we anticipate substantial improvements with optimization of sample and experimental conditions. In both spectra, peaks from Ala, Ile, Ser, and Thr populate the regions expected for β-sheet conformation. For example, Ail has four Thr, and four signals are observed with ^{13}C shifts in the expected region (Fig. 6a).

Solid-State NMR Studies of Ail and OmpX in Planar Lipid Bilayers. OS solid-state NMR spectra yield single-line resonances that directly reflect the orientations of molecular sites relative to the membrane (Fig. 6d–f). Multidimensional SLF and heteronuclear correlation experiments [78–83] with uniformly and selectively labeled samples can be used to resolve the spectra and measure orientation-dependent DC and CSA frequencies for structure determination [39, 40]. Both the one-dimensional ^{15}N spectra and the two-dimensional ^{1}H/^{15}N SLF spectra of OmpX in magnetically aligned DMPC/DHPC bilayers show very high resolution (Fig. 6d–f). They benefit significantly from the sensitivity and resolution enhancements possible through the use of high magnetic fields and newly developed radiofrequency probes [84]. These spectra were assigned using SASR/AssignFit methods (Fig. 5). Additional assignments can be obtained using a combination of selective isotope labeling schemes with structural model fitting [25, 50–52, 55, 85–87], comparisons with isotropic NMR data [88], and multidimensional triple-resonance experiments [89–93], similar to the multidimensional ^{15}N spin-exchange experiments that we developed for resonance assignments in a helical membrane protein [56]; we anticipate that the latter will be even more useful for β-barrel proteins because of the wider chemical shift dispersion from neighboring amide sites available in their spectra.

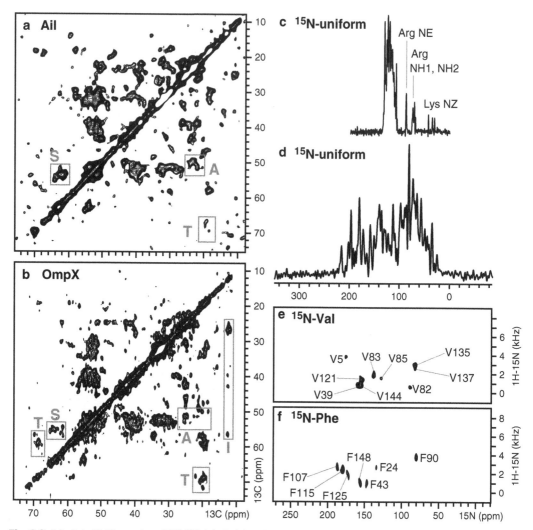

Fig. 6 Solid-state NMR spectra of $^{15}N/^{13}C$-labeled OmpX and Ail in lipid bilayers. (**a**, **b**) Two-dimensional $^{13}C/^{13}C$ correlation MAS solid-state NMR spectra of Ail and OmpX in DMPC proteoliposomes. (**c**) One-dimensional ^{15}N MAS solid-state NMR spectrum of OmpX in DMPC proteoliposomes. (**d**) One-dimensional ^{15}N OS solid-state NMR spectrum of OmpX in parallel (bilayer normal ∥ magnetic field) magnetically aligned DMPC/DHPC lipid bilayers. (**e**, **f**) Two-dimensional $^{1}H/^{15}N$ SLF OS solid-state NMR spectra of OmpX in parallel magnetically aligned DMPC/DHPC bilayers. Proteins are uniformly $^{15}N/^{13}C$ labeled (**a**–**c**), uniformly ^{15}N labeled (**d**), ^{15}N Val labeled (**c**), or ^{15}N Phe labeled (**d**)

4   Conclusions

Recent progress in sample optimization, instrumentation, and NMR experiments enables very high-resolution solid-state NMR spectra to be obtained for membrane proteins in lipid bilayers. For example, two bacterial outer membrane proteins, OmpX and Ail, refolded in lipid bilayers yield very high-quality MAS and OS solid-state NMR spectra, where several individual peaks can be resolved and assigned (Fig. 6). The resolution observed even in one-dimensional spectra is remarkable, indicating that structure determination of these β-barrel membrane proteins in natural lipid bilayer environments is within reach.

Acknowledgments

This research was supported by grants from the National Institutes of Health (R21 GM075917; R21 GM094727; R01 GM100265). The NMR studies utilized the NMR Facility at Sanford-Burnham Medical Research Institute, and the Resource for Molecular Imaging of Proteins at UCSD, each supported by grants from the National Institutes of Health (P30 CA030199; P41 EB002031).

References

1. White SH, Wiener MC (1994) Determination of the structure of fluid lipid bilayer membranes. In: Disalvo EA, Simon SA (eds) Permeability and stability of lipid bilayers. CRC, Boca Raton, FL, pp 1–19

2. Marrink S, Berkowitz M (1995) Water and membranes. In: Disalvo EA, Simon SA (eds) Permeability and stability of lipid bilayers. CRC, Boca Raton, FL, pp 21–48

3. Engelman DM (1996) Crossing the hydrophobic barrier: insertion of membrane proteins. Science 274:1850–1851

4. de Planque MR, Goormaghtigh E, Greathouse DV, Koeppe RE 2nd, Kruijtzer JA, Liskamp RM, de Kruijff B, Killian JA (2001) Sensitivity of single membrane-spanning alpha-helical peptides to hydrophobic mismatch with a lipid bilayer: effects on backbone structure, orientation, and extent of membrane incorporation. Biochemistry 40:5000–5010

5. Page RC, Li C, Hu J, Gao FP, Cross TA (2007) Lipid bilayers: an essential environment for the understanding of membrane proteins. Magn Reson Chem 45:S2–S11

6. White SH (2009) Biophysical dissection of membrane proteins. Nature 459:344–346

7. Cross TA, Sharma M, Yi M, Zhou HX (2011) Influence of solubilizing environments on membrane protein structures. Trends Biochem Sci 36:117–125

8. Anfinsen CB (1973) Principles that govern the folding of protein chains. Science 181:223–230

9. Gluck JM, Wittlich M, Feuerstein S, Hoffmann S, Willbold D, Koenig BW (2009) Integral membrane proteins in nanodiscs can be studied by solution NMR spectroscopy. J Am Chem Soc 131:12060–12061

10. Raschle T, Hiller S, Yu TY, Rice AJ, Walz T, Wagner G (2009) Structural and functional characterization of the integral membrane protein VDAC-1 in lipid bilayer nanodiscs. J Am Chem Soc 131:17777–17779

11. Shenkarev ZO, Lyukmanova EN, Paramonov AS, Shingarova LN, Chupin VV, Kirpichnikov MP, Blommers MJ, Arseniev AS (2010) Lipid-protein nanodiscs as reference medium in detergent screening for high-resolution NMR studies of integral membrane proteins. J Am Chem Soc 132:5628–5629

12. Warschawski D.E. http://wwwdrorlistcom/nmrhtml.

13. White S. http://blancobiomoluciedu/mpstruc/listAll/list.

14. Cornelis GR (2000) Molecular and cell biology aspects of plague. Proc Natl Acad Sci USA 97:8778–8783

15. Bartra SS, Styer KL, O'Bryant DM, Nilles ML, Hinnebusch BJ, Aballay A, Plano GV (2008) Resistance of yersinia pestis to complement-dependent killing is mediated by the Ail outer membrane protein. Infect Immun 76:612–622

16. Tsang TM, Felek S, Krukonis ES (2010) Ail binding to fibronectin facilitates Yersinia pestis binding to host cells and Yop delivery. Infect Immun 78:3358–3368

17. Vogt J, Schulz GE (1999) The structure of the outer membrane protein OmpX from Escherichia coli reveals possible mechanisms of virulence. Structure 7:1301–1309

18. Fernandez C, Hilty C, Wider G, Guntert P, Wuthrich K (2004) NMR structure of the integral membrane protein OmpX. J Mol Biol 336:1211–1221

19. Yamashita S, Lukacik P, Barnard TJ, Noinaj N, Felek S, Tsang TM, Krukonis ES, Hinnebusch BJ, Buchanan SK (2011) Structural insights into Ail-mediated adhesion in Yersinia pestis. Structure 19:1672–1682

20. Plesniak LA, Mahalakshmi R, Rypien C, Yang Y, Racic J, Marassi FM (2011) Expression, refolding, and initial structural characterization of the Y. pestis Ail outer membrane protein in lipids. Biochim Biophys Acta 1808:482–489

21. Mahalakshmi R, Franzin CM, Choi J, Marassi FM (2007) NMR structural studies of the bacterial outer membrane protein OmpX in oriented lipid bilayer membranes. Biochim Biophys Acta 1768:3216–3224

22. Mahalakshmi R, Marassi FM (2008) Orientation of the Escherichia coli outer membrane protein OmpX in phospholipid bilayer membranes determined by solid-state NMR. Biochemistry 47:6531–6538

23. Delaglio F, Kontaxis G, Bax A (2000) Protein structure determination using molecular fragment replacement and NMR dipolar couplings. J Am Chem Soc 122:2142–2143

24. Tian F, Valafar H, Prestegard JH (2001) A dipolar coupling based strategy for simultaneous resonance assignment and structure determination of protein backbones. J Am Chem Soc 123:11791–11796

25. Marassi FM, Opella SJ (2003) Simultaneous assignment and structure determination of a membrane protein from NMR orientational restraints. Protein Sci 12:403–411

26. Cavalli A, Salvatella X, Dobson CM, Vendruscolo M (2007) Protein structure determination from NMR chemical shifts. Proc Natl Acad Sci USA 104:9615–9620

27. Shen Y, Lange O, Delaglio F, Rossi P, Aramini JM, Liu G, Eletsky A, Wu Y, Singarapu KK, Lemak A, Ignatchenko A, Arrowsmith CH, Szyperski T, Montelione GT, Baker D, Bax A (2008) Consistent blind protein structure generation from NMR chemical shift data. Proc Natl Acad Sci USA 105:4685–4690

28. Wishart DS, Arndt D, Berjanskii M, Tang P, Zhou J, Lin G (2008) CS23D: a web server for rapid protein structure generation using NMR chemical shifts and sequence data. Nucleic Acids Res 36:W496–W502

29. Rohl CA, Strauss CE, Misura KM, Baker D (2004) Protein structure prediction using Rosetta. Methods Enzymol 383:66–93

30. Das R, Baker D (2008) Macromolecular modeling with rosetta. Annu Rev Biochem 77:363–382

31. Raman S, Lange OF, Rossi P, Tyka M, Wang X, Aramini J, Liu G, Ramelot TA, Eletsky A, Szyperski T, Kennedy MA, Prestegard J, Montelione GT, Baker D (2010) NMR structure determination for larger proteins using backbone-only data. Science 327:1014–1018

32. Berardi MJ, Shih WM, Harrison SC, Chou JJ (2011) Mitochondrial uncoupling protein 2 structure determined by NMR molecular fragment searching. Nature 476:109–113

33. Das BB, Nothnagel HJ, Lu GJ, Son WS, Tian Y, Marassi FM, Opella SJ (2012) Structure determination of a membrane protein in proteoliposomes. J Am Chem Soc 134: 2047–2056

34. Sharma M, Yi M, Dong H, Qin H, Peterson E, Busath DD, Zhou HX, Cross TA (2010) Insight into the mechanism of the influenza A proton channel from a structure in a lipid bilayer. Science 330:509–512

35. Griffin RG (1998) Dipolar recoupling in MAS spectra of biological solids. Nat Struct Biol 5(Suppl):508–512

36. Luca S, Heise H, Baldus M (2003) High-resolution solid-state NMR applied to polypeptides and membrane proteins. Acc Chem Res 36:858–865

37. McDermott A (2009) Structure and dynamics of membrane proteins by magic angle spinning solid-state NMR. Annu Rev Biophys 38: 385–403

38. Cross TA, Quine JR (2000) Protein structure in anisotropic environments: development of orientational constraints. Develop Orien Constr 12:55–70

39. Marassi FM (2002) NMR of peptides and proteins in membranes. J Magn Reson 14: 212–224

40. Opella SJ, Marassi FM (2004) Structure determination of membrane proteins by NMR spectroscopy. Chem Rev 104:3587–3606

41. Marassi FM, Das BB, Lu GJ, Nothnagel HJ, Park SH, Son WS, Tian Y, Opella SJ (2011) Structure determination of membrane proteins in five easy pieces. Methods 55:363–369

42. Singer SJ, Nicholson GL (1972) The fluid mosaic model of the structure of cell membranes. Science 175:720–731

43. Edidin M (1974) Rotational and translational diffusion in membranes. Annu Rev Biophys Bioeng 3:179–201

44. Cherry RJ (1975) Protein mobility in membranes. FEBS Lett 55:1–7

45. Saffman PG, Delbruck M (1975) Brownian motion in biological membranes. Proc Natl Acad Sci USA 72:3111–3113

46. Cherry RJ, Muller U, Schneider G (1977) Rotational diffusion of bacteriorhodopsin in lipid membranes. FEBS Lett 80:465–469

47. Kovacs FA, Cross TA (1997) Transmembrane four-helix bundle of influenza a M2 protein channel: structural implications from helix tilt and orientation. Biophys J 73:2511–2517

48. Tian F, Song Z, Cross TA (1998) Orientational constraints derived from hydrated powder samples by two-dimensional PISEMA. J Magn Reson 135:227–231

49. Cady SD, Goodman C, Tatko CD, DeGrado WF, Hong M (2007) Determining the orientation of uniaxially rotating membrane proteins using unoriented samples: a 2H, 13C, AND 15N solid-state NMR investigation of the dynamics and orientation of a transmembrane helical bundle. J Am Chem Soc 129:5719–5729

50. Marassi FM, Opella SJ (2000) A solid-state NMR index of helical membrane protein structure and topology. J Magn Reson 144:150–155

51. Wang J, Denny J, Tian C, Kim S, Mo Y, Kovacs F, Song Z, Nishimura K, Gan Z, Fu R, Quine JR, Cross TA (2000) Imaging membrane protein helical wheels. J Magn Reson 144: 162–167

52. Marassi FM (2001) A simple approach to membrane protein secondary structure and topology based on NMR spectroscopy. Biophys J 80:994–1003

53. Nishimura K, Kim S, Zhang L, Cross TA (2002) The closed state of a H + channel helical bundle combining precise orientational and distance restraints from solid state NMR. Biochemistry 41:13170–13177

54. Park SH, Das BB, Casagrande F, Tian Y, Nothnagel HJ, Chu M, Kiefer H, Maier K, De Angelis A, Marassi FM, Opella SJ (2012) Structure of the chemokine receptor CXCR1 in phospholipid bilayers. Biophys J 102(3):422

55. Asbury T, Quine JR, Achuthan S, Hu J, Chapman MS, Cross TA, Bertram R (2006) PIPATH: an optimized algorithm for generating alpha-helical structures from PISEMA data. J Magn Reson 183:87–95

56. Marassi FM, Gesell JJ, Valente AP, Kim Y, Oblatt-Montal M, Montal M, Opella SJ (1999) Dilute spin-exchange assignment of solid-state NMR spectra of oriented proteins: acetylcholine M2 in bilayers. J Biomol NMR 14:141–148

57. Marassi FM, Opella SJ (2002) Using pisa pies to resolve ambiguities in angular constraints from PISEMA spectra of aligned proteins. J Biomol NMR 23:239–242

58. Tian Y, Schwieters CD, Opella SJ, Marassi FM (2012) AssignFit: a program for simultaneous assignment and structure refinement from solid-state NMR spectra. J Magn Reson 214:42–50

59. Wishart DS, Bigam CG, Yao J, Abildgaard F, Dyson HJ, Oldfield E, Markley JL, Sykes BD (1995) 1H, 13C and 15N chemical shift referencing in biomolecular NMR. J Biomol NMR 6:135–140

60. Morcombe CR, Zilm KW (2003) Chemical shift referencing in MAS solid state NMR. J Magn Reson 162:479–486

61. Delaglio F, Grzesiek S, Vuister GW, Zhu G, Pfeifer J, Bax A (1995) NMRPipe: a multidimensional spectral processing system based on UNIX pipes. J Biomol NMR 6:277–293

62. Goddard TD, Kneller DG (2004) SPARKY 3. University of California, San Francisco

63. Xhao X, Eden M, Levitt MH (2001) Recoupling of heteronuclear dipolar interactions in solid-state NMR using symmetry-based pulse sequences. J Phys Chem B 342:353–361

64. Castellani F, van Rossum B, Diehl A, Schubert M, Rehbein K, Oschkinat H (2002) Structure of a protein determined by solid-state magic-angle-spinning NMR spectroscopy. Nature 420:98–102

65. Watts A, Straus SK, Grage SL, Kamihira M, Lam YH, Zhao X (2004) Membrane protein structure determination using solid-state NMR. Methods Mol Biol 278:403–473

66. Franks WT, Wylie BJ, Schmidt HL, Nieuwkoop AJ, Mayrhofer RM, Shah GJ, Graesser DT, Rienstra CM (2008) Dipole tensor-based atomic-resolution structure determination of a nanocrystalline protein by solid-state NMR. Proc Natl Acad Sci USA 105:4621–4626

67. Cady SD, Schmidt-Rohr K, Wang J, Soto CS, Degrado WF, Hong M (2010) Structure of the amantadine binding site of influenza M2 proton channels in lipid bilayers. Nature 463:689–692

68. Tycko R (2011) Solid-state NMR studies of amyloid fibril structure. Annu Rev Phys Chem 62:279–299

69. McDermott A, Polenova T, Bockmann A, Zilm KW, Paulson EK, Martin RW, Montelione GT (2000) Partial NMR assignments for uniformly (13C, 15N)-enriched BPTI in the solid state. J Biomol NMR 16:209–219

70. Bockmann A, Lange A, Galinier A, Luca S, Giraud N, Juy M, Heise H, Montserret R, Penin F, Baldus M (2003) Solid state NMR sequential resonance assignments and conformational analysis of the 2x10.4 kDa dimeric form of the bacillus subtilis protein Crh. J Biomol NMR 27:323–339

71. Igumenova TI, Wand AJ, McDermott AE (2004) Assignment of the backbone resonances for microcrystalline ubiquitin. J Am Chem Soc 126:5323–5331

72. Igumenova TI, McDermott AE, Zilm KW, Martin RW, Paulson EK, Wand AJ (2004) Assignments of carbon NMR resonances for microcrystalline ubiquitin. J Am Chem Soc 126:6720–6727

73. Marulanda D, Tasayco ML, McDermott A, Cataldi M, Arriaran V, Polenova T (2004) Magic angle spinning solid-state NMR spectroscopy for structural studies of protein interfaces. Resonance assignments of differentially enriched escherichia coli thioredoxin reassembled by fragment complementation. J Am Chem Soc 126:16608–16620

74. Mills FD, Antharam VC, Ganesh OK, Elliott DW, McNeill SA, Long JR (2008) The helical structure of surfactant peptide KL4 when bound to POPC: POPG lipid vesicles. Biochemistry 47:8292–8300

75. Szeverenyi NM, Sullivan MJ, Maciel GE (1982) Observation of spin exchange by two-dimensional fourier transform 13C cross-polarization-magic angle spinning. J Magn Reson 47:462–475

76. Frey MH, Opella SJ (1984) 13C Spin exchange in amino acids and peptides. J Am Chem Soc 106:4942–4945

77. Baldus M, Geurts DG, Meier BH (1998) Broadband dipolar recoupling in rotating solids: a numerical comparison of some pulse schemes. Solid State Nucl Magn Reson 11:157–168

78. Wu CH, Ramamoorthy A, Opella SJ (1994) High-resolution heteronuclear dipolar solid-state NMR spectroscopy. J Magn Reson 109:270–272

79. Ramamoorthy A, Wu CH, Opella SJ (1995) Three-dimensional solid-state NMR experiment that correlates the chemical shift and dipolar coupling frequencies of two heteronuclei. J Magn Reson B 107:88–90

80. Ramamoorthy A, Opella SJ (1995) Two-dimensional chemical shift/heteronuclear dipolar coupling spectra obtained with polarization inversion spin exchange at the magic angle and magic-angle sample spinning (PISEMAMAS). Solid State Nucl Magn Reson 4:387–392

81. Ramamoorthy A, Marassi FM, Zasloff M, Opella SJ (1995) Three-dimensional solid-state NMR spectroscopy of a peptide oriented in membrane bilayers. J Biomol NMR 6:329–334

82. Nevzorov AA, Opella SJ (2003) A "magic sandwich" pulse sequence with reduced offset dependence for high-resolution separated local field spectroscopy. J Magn Reson 164:182–186

83. Dvinskikh SV, Yamamoto K, Ramamoorthy A (2006) Heteronuclear isotropic mixing separated local field NMR spectroscopy. J Chem Phys 125:34507

84. Grant CV, Wu CH, Opella SJ (2010) Probes for high field solid-state NMR of lossy biological samples. J Magn Reson 204:180–188

85. Nevzorov AA, Opella SJ (2003) Structural fitting of PISEMA spectra of aligned proteins. J Magn Reson 160:33–39

86. Franzin CM, Teriete P, Marassi FM (2007) Structural similarity of a membrane protein in micelles and membranes. J Am Chem Soc 129:8078–8079

87. Tian Y, Schwieters CD, Opella SJ, Marassi FM (2011) AssignFit: a program for simultaneous assignment and structure refinement from solid-state NMR spectra. J Magn Reson 214(1):42–50

88. De Angelis AA, Howell SC, Opella SJ (2006) Assigning solid-state NMR spectra of aligned proteins using isotropic chemical shifts. J Magn Reson 183:329–332

89. Ramamoorthy A, Gierasch LM, Opella SJ (1996) Three-dimensional solid-state NMR correlation experiment with 1H homonuclear spin exchange. J Magn Reson B 111: 81–84

90. Nevzorov AA (2008) Mismatched Hartmann-Hahn conditions cause proton-mediated inter-molecular magnetization transfer between dilute low-spin nuclei in NMR of static solids. J Am Chem Soc 130:11282–11283

91. Nevzorov AA (2009) High-resolution local field spectroscopy with internuclear correlations. J Magn Reson 201:111–114

92. Knox RW, Lu GJ, Opella SJ, Nevzorov AA (2010) A resonance assignment method for oriented-sample solid-state NMR of proteins. J Am Chem Soc 132:8255–8257

93. Xu J, Smith PE, Soong R, Ramamoorthy A (2011) A proton spin diffusion based solid-state NMR approach for structural studies on aligned samples. J Phys Chem B 115: 4863–4871

94. Ketchem RR, Hu W, Cross TA (1993) High-resolution conformation of gramicidin a in a lipid bilayer by solid-state NMR. Science 261:1457–1460

95. Opella SJ, Marassi FM, Gesell JJ, Valente AP, Kim Y, Oblatt-Montal M, Montal M (1999) Structures of the M2 channel-lining segments from nicotinic acetylcholine and NMDA receptors by NMR spectroscopy. Nat Struct Biol 6:374–379

Chapter 9

On the Role of NMR Spectroscopy for Characterization of Antimicrobial Peptides

Fernando Porcelli, Ayyalusamy Ramamoorthy, George Barany, and Gianluigi Veglia

Abstract

Antimicrobial peptides (AMPs) provide a primordial source of immunity, conferring upon eukaryotic cells resistance against bacteria, protozoa, and viruses. Despite a few examples of anionic peptides, AMPs are usually relatively short positively charged polypeptides, consisting of a dozen to about a hundred amino acids, and exhibiting amphipathic character. Despite significant differences in their primary and secondary structures, all AMPs discovered to date share the ability to interact with cellular membranes, thereby affecting bilayer stability, disrupting membrane organization, and/or forming well-defined pores. AMPs selectively target infectious agents without being susceptible to any of the common pathways by which these acquire resistance, thereby making AMPs prime candidates to provide therapeutic alternatives to conventional drugs. However, the mechanisms of AMP actions are still a matter of intense debate. The structure–function paradigm suggests that a better understanding of how AMPs elicit their biological functions could result from atomic resolution studies of peptide–lipid interactions. In contrast, more strict thermodynamic views preclude any roles for three-dimensional structures. Indeed, the design of selective AMPs based solely on structural parameters has been challenging. In this chapter, we will focus on selected AMPs for which studies on the corresponding AMP–lipid interactions have helped reach an understanding of how AMP effects are mediated. We will emphasize the roles of both liquid- and solid-state NMR spectroscopy for elucidating the mechanisms of action of AMPs.

Key words Antimicrobial peptides, Solution NMR, Solid-state NMR, Lipid membranes

1 Introduction

The increasing resistance of bacteria to conventional antibiotics represents a serious global emergency for human health that can only be addressed by the discovery and clinical implementation of new antimicrobial drugs [1–8]. Over the past 40 years, however, only three new classes of antibiotics have been found that could be used as drugs: lipopeptides, oxazolidinones, and streptogramins [9]. Since antimicrobial peptides (AMPs) are not susceptible to any of the common mechanisms whereby organisms acquire drug

Giovanna Ghirlanda and Alessandro Senes (eds.), *Membrane Proteins: Folding, Association, and Design*,
Methods in Molecular Biology, vol. 1063, DOI 10.1007/978-1-62703-583-5_9, © Springer Science+Business Media, LLC 2013

resistance, representatives of the class of AMPs offer an alternative to standard antimicrobial therapy.

AMPs constitute one of the first evolved chemical defense mechanisms of eukaryotic cells against bacteria, protozoa, fungi, parasites, and viruses [10–12]. AMPs have been isolated from a variety of organisms, including humans, and are present in virtually all living species from bacteria to invertebrates to vertebrates [13]. The first AMP, nisin, was isolated in 1947 [14, 15], but systematic research on AMPs only started in the early 1980s, upon publication of Hans Boman's paper on the isolation, purification, sequencing, and specificity of two antibacterial peptides, cecropins A and B, involved in insect immunity [10]. Since this pioneering research, more than 1,800 AMPs have been identified and isolated from several organisms, and structural analogues have been synthesized and tested for clinical purposes [16–22]. A comprehensive database of isolated or synthesized AMPs is available at http://aps.unmc.edu/AP/main.php [22].

Properties of AMPs in their role as antibiotics include a broad spectrum of antibacterial activity, high selectivity, and the ability to disrupt bacterial cell membranes [2, 11, 23, 24]. In addition, several AMPs can modulate immune and inflammatory responses, inducing phagocytosis, promotion of immune cell recruitment, regulation of angiogenesis, and stimulation of prostaglandin release [13, 19, 20, 25–29].

AMPs can be grouped into diverse classes based on the pathways of their biosynthesis, their structural properties, and their biological activities. AMPs are either *ribosomally synthesized* oligopeptides, when produced by mammals, birds, amphibians, insects, plants, or certain microorganisms [30, 31], or *non-ribosomally synthesized* peptides when produced by bacteria and fungi, i.e., bacteriocins. Additionally, AMPs can be subdivided as antiviral, antifungal, anticancer, antiparasital, insecticidal, spermicidal, anti-HIV, and/or as having chemotactic activity [22]. Finally, AMPs can be classified on the basis of molecular structure as (a) linear amphipathic α-helical, e.g., cathelicidins, magainins, and cecropins [32–38]; (b) amphiphilic β-sheet structures containing disulphide bonds, e.g., defensins [39–41]; and (c) turns and extended structures, e.g., protegrins [42–45].

AMPs are usually cationic, relatively short (less than a hundred amino acid residues), and amphipathic, with approximately half of the residues hydrophobic so as to enable their associations with cell membranes [13, 46–54]. In fact, to elicit their biological functions, it is necessary for AMPs to interact with the bacterial cytoplasmic membrane, under which conditions AMPs often adopt a secondary structure [48, 54–58]. From a physical chemical standpoint, antibacterial activity of AMPs seems to be related to amino acid composition, which may also dictate their selectivity toward bacterial cell membranes with specific lipid composition and structure [59, 60].

However, a unified mechanism of action for AMPs has not been identified. It is possible that each class of AMPs has a specific mechanism of action.

2 Proposed Mechanisms of Action

The biological actions of AMPs are not well understood. Researchers agree that the primary target for many AMPs is the outer cell membrane [61–64]. Two main membrane interaction mechanisms have been proposed: (a) pore formation across the membrane and (b) a carpet-like mechanism [65, 66]. These two models share some similarities and involve four major steps: attraction, attachment, insertion, and membrane permeation [13, 50]. The initial *attraction* of the peptide toward the membrane is driven by electrostatic interactions between the positively charged residues (generally Arg and Lys) and the negative lipid headgroups of the bacterial membrane [67–69]. The *attachment* step depends on the distribution of hydrophobic and hydrophilic residues of AMPs. After reaching the membrane, AMPs pass a concentration threshold after which cell membrane permeability is altered [48, 62, 70–73]. At low peptide/lipid ratio (P/L) ratio, AMPs are positioned with their backbones parallel to the surface of the lipid bilayer, so that the AMPs interact with the lipid headgroups. With this topology, AMPs often insert into the membrane, modifying its thickness and curvature [74–80]. As the P/L ratio increases, AMPs orient perpendicularly to the membrane and insert into the membrane bilayer while forming pores [81]. Amino acid composition, charge, and amphipathicity drive AMP *insertion*. In addition, several studies show that membrane partitioning causes AMPs to adopt a regular secondary structure; this is referred to as the *folding-upon-binding* mechanism [82, 83]. The latter effect causes a substantial reduction of the free energy of insertion, due to the formation of intramolecular hydrogen bonds [82, 84, 85]. Within the lipid bilayer, AMPs can assume different architectures, including *toroidal, carpet-like,* or *barrel-stave* [65, 86–88]. In the *toroidal model,* the polar faces of AMPs associate with lipid polar headgroups and insert into the membrane in a manner that involves induced bending of the lipid monolayers and concomitant pore formation [50, 89]. In the *carpet-like* model, AMPs coat the bilayer outer surface, just like a carpet; after reaching a certain concentration threshold, the AMPs disrupt the membrane in a detergent-like manner, causing membrane lysis [38, 50, 87, 90]. In the *barrel-stave* model exemplified by alamethicin, AMPs aggregate to form a bundle of monomers (staves in a barrel), forming a pore in the membrane by thinning and bending of the inner and outer membrane leaflets [66, 91].

Disruption of the membrane dissipates the electrochemical gradients necessary for cells to thrive, while also causing the loss of essential cytoplasmic constituents. It has been reported that the relative preference of barrel-stave vs. toroidal mechanisms depends on peptide length [57, 75, 92–94]. Modifications of the peptide sequence involving changes in charge distribution, hydrophobicities, and surface areas of different amino acid side chains have been shown to modulate the specificities and mechanisms of action for AMPs [49, 53, 70, 95, 96].

Over the past few years, a more thermodynamic approach has been applied to explain membrane permeabilization by AMPs. This viewpoint is based on the efflux of fluorescent dye observed with lipid vesicles and precludes any structural factors [48, 62, 89, 97–101]. Such studies suggest two possible mechanisms for cell permeation: (a) *graded* or (b) *all-or-none* [89, 101–104]. In the *graded* mode, all vesicles release part of their content, while in the *all-or-none* mode only a fraction of the vesicles release their contents while the remainder do not lose any dye. In essence, the *graded* mechanism can be understood in terms of a destabilizing detergent-like effect on all vesicles, whereas the *all-or-none* mode of depletion suggests the formation of multimeric pores or the disruption of the vesicle structure [89, 101–104].

To what extent do structures affect the activities and selectivities of AMPs? Is it possible that millions of years of evolution did not encode specificity and function into these peptides as for larger proteins? At this writing, the questions just posed remain unanswered. While there are clear examples of structure–function relationship for several classes of AMPs, the proponents of a more thermodynamic viewpoint invoke the many examples where AMPs with all D-amino acid residues or scrambled sequences still manifest antimicrobial activity. In favor of the structure–activity relationship is the possibility that the landscape of the AMPs structural propensities, as well as their interactions with lipid membranes, is still limited to selected cases.

3   Bacterial Membrane Composition and Membrane Mimicking Systems for Structural Studies

The plasma membrane contains both neutral and zwitterionic lipids, such as phosphatidylcholine (PC), or acidic lipids, such as phosphatidylserine (PS), cardiolipin, and phosphatidylglycerol (PG) [65, 105, 106]. While it is relatively easy to reproduce the relevant lipid composition with synthetic lipids, it is very challenging to mimic the complexity of cell membrane, a significant concern for both functional and structural studies. For structural work several different model systems have been employed. As a general guideline, the membrane mimetic systems should not only represent the natural environment as closely as possible, but should also be compatible with structural techniques such as NMR, CD, and fluorescence measurements. Originally, NMR researchers used aqueous solutions of organic solvents such as trifluoroethanol (TFE) or hexafluoroisopropanol (HFIP). Under such conditions, most known AMPs are forced to adopt α-helical conformations, which often do not correlate with biological activities [107–110]. To overcome this problem, detergent micelles, such as those formed from DPC (dodecylphosphocholine) plus SDS (sodium dodecylsulfate), have been used in order to better simulate the membrane interface. Micelles are definitely a superior medium to study AMPs, since they provide a water/lipid interface similar to the lipid membranes.

Usually, micelles are spherical monolayers with a diameter of ~3 nm that can assume elliptical or rod-like shapes, depending on detergent concentration and chain length [111, 112]. Their small size and fast tumbling enable solution NMR spectroscopy of micelle-bound AMPs and small membrane proteins [34, 113–117]. Detergent micelles, however, are only a rough approximation of a membrane bilayer. The monolayer structure and small curvature radius of micelles can cause peptides and proteins to adopt incorrectly folded conformations or to aggregate [118]. Recently, discoidal micelles or bicelles have been used in NMR spectroscopy of both membrane peptides and proteins [112, 118–122]. Generally, bicelles are formed by long-chain phospholipids (DMPC, DMPG) and amphiphilic molecules, such as CHAPSO, or a short-chain lipid, such as DHPC. The structures and shapes of bicelles depend on the lipid to detergent ratio (called q). At high detergent concentrations ($q \sim 0.1$–0.8), bicelles assume a discoidal shape and tumble rapidly in solution. At low detergent concentrations ($q \sim 2.8$–6), bicelles may resemble perforated bilayer sheets [123–125]. While AMPs and membrane proteins can be analyzed by solution NMR techniques in isotropic bicelles, anisotropic bicelles align spontaneously in magnetic fields, enabling peptides and proteins to be analyzed using oriented solid-state NMR approaches [126–130]. In addition to magnetically oriented bicelles, AMPs can be reconstituted in mechanically aligned membrane bilayers supported on glass plates [131–133]. This method of sample preparation, originally developed by Seelig and co-workers [134–136], has been used widely to analyze membrane peptides and proteins [133, 137–142]. This approach, however, is quite laborious and prone to artifacts since AMPs have a tendency to aggregate and disrupt the organization of lipid bilayers, making it difficult to discern the mechanism of action. Therefore, many studies are being carried out in lipid vesicles, which are amenable to magic angle spinning (MAS) NMR techniques [143, 144].

4 NMR Spectroscopic Approaches to Study AMPs

Classical solution NMR techniques have been used successfully to determine three-dimensional structures of bioactive peptides in mixtures of water with fluorinated organic solvents [107, 110, 145, 146], in detergent micelles [34, 38, 115, 147, 148], and in isotropic bicelles [119, 149, 150]. NOESY-based techniques have been used to measure distance restraints for structure calculations. In the case of recombinant-expressed peptides, weak alignment of peptides using acrylamide gels has been used to obtain residual dipolar couplings for orientational-dependent restraints [151–153]. The positioning of peptides with respect to micelle surfaces has been inferred by paramagnetic relaxation enhancement mapping [34, 38, 147, 154, 155], water exposure measurements based on exchange peaks with the water signal [156–158], detection of NOEs between protonated micelles and peptides [159–162], or saturation transfer techniques [163–165].

Despite the results from solution NMR, accurate descriptions of interactions between any given peptide and the membrane within which it exerts bioactivity require the use of lipid membranes [35, 70, 128, 129, 131, 141, 142, 166–177]. Meeting this need, solid-state NMR techniques are well suited for studying membrane-embedded peptides and proteins. The two major approaches are (a) static or oriented solid-state NMR [170, 178–182], wherein nuclear anisotropic interactions are obtained from aligning samples with respect to the direction of the static magnetic field, and (b) MAS NMR, whereby samples are spun at the magic angle ($\theta \sim 54.7°$) to remove the effects of chemical shift anisotropy and dipolar couplings [70, 173, 183–185]. With fast spinning at the magic angle, resonances can reach line widths similar to those observed in solution NMR spectra.

In the following, we will focus on selected AMPs (magainins, pardaxins, distinctin, and cathelicidins) that our group has studied, and in so doing, highlight the role of NMR in understanding structure–function relationships.

4.1 Magainins

This class of AMPs is expressed by amphibians as a defense against microbes. Magainins were first identified by Zasloff in the early 1980s [36, 186]. In general, magainins are helical and amphipathic and exhibit a broad spectrum of antimicrobial activities [33, 34, 86, 107, 108, 172]. Their structures and orientations with respect to membrane bilayers have been studied extensively, as has been reviewed by the Zasloff, Opella, and Bechinger groups [35, 108, 172, 187–192]. In our laboratory, we determined the high-resolution structures and lipid interactions of two synthetic variants of magainins (MSI-78 and MSI-594) that were originally designed by Genaera Corporation for clinical use [193].

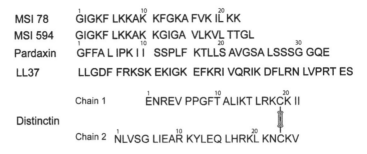

Fig. 1 Primary sequences of the helical AMPs studied by our groups

MSI-78, an analog of magainin 2, is a 22-residue polypeptide known as pexiganan that is being tested in phase II and III clinical trials for treatment of diabetic foot ulcers. MSI-78 interacts strongly with model bacterial cell membranes, selectively inducing bacterial membrane disruption while leaving mammalian cells essentially unperturbed [194]. MSI-594 is a 24-residue peptide with high efficacy against herpes simplex virus I [195]. The primary sequences of MSI-78 and MSI-594 differ significantly only at the C-terminus (Fig. 1). In DPC micelles, each peptide gives a high-resolution spectral fingerprint that enables the sequential assignment of nearly all of the backbone and side chain resonances. In DPC micelles, MSI-78 assumes a distorted α-helical conformation throughout the entire length of the polypeptide chain. A significant feature of MSI-78 is its ability to form stable antiparallel dimers, as defined by head to tail NOEs contacts. The latter may explain its resistance to proteolysis and its higher efficacy with respect to MSI-594. As it turns out, MSI-594 adopts a well-defined helical conformation, but does not dimerize, a property that may account for the significantly lower antibacterial activity displayed by this molecule. Dimerization is a common feature for the majority of magainins, as exemplified by magainin 2, which forms stable dimers in PC vesicles [196].

MSI-78 dimerization seems to be encoded in its primary sequence (Fig. 1). In fact, the central hydrophobic core of the MSI-78 dimer is formed by a "phenylalanine zipper" that hold the two protomers together, with charged residues (e.g., lysines) pointing toward the bulk solvent. In contrast, MSI-594 lacks phenylalanine residues in position 13 and 16, preventing formation of a stable dimer. Moreover, the GIG motif located in the middle of the primary sequence enables MSI-78 to assume a curved structure, thereby inducing conformational dynamics (reported by broader NMR lines) that further hampers the dimerization process. The Ramamoorthy laboratory performed MAS solid-state NMR experiments on the magainin analogs reconstituted in POPC and 3:1 POPC/POPG multilamellar vesicles (MLVs).

Peptides that had been ^{13}C-labeled at the α-carbonyl displayed isotropic chemical shift values of ~176–179 ppm that are characteristic of α-helical conformations [197] and confirmatory of the helical structure determined by solution NMR. These authors also found that the line widths of MSI-594 are broader than the corresponding resonances of MSI-78, signifying that conformational exchange dynamics are more accentuated for MSI-78. Again, solution and solid-state NMR data were consistent with each other, pointing toward the same conclusions.

Based on NMR studies both in micelles and in the solid-state, we proposed that differences observed for MSI-78 and MSI-594, in terms of both function and selectivity, may be due to their differing tendencies to self-associate. These results support the in vivo antimicrobial activity of MSI-78, which has been proposed to oligomerize into toroidal-type structures that permeate bacterial cells. In contrast, the behavior of MSI-594 is completely different, suggesting an alternative mechanism of action that must await further studies for elucidation [34].

4.2 Pardaxins

This class of small shark-repellent AMPs are isolated from the sole fish of genus *Pardachirus* [198, 199]. Pardaxins interact with cell membranes, causing disruption of ionic transport and presynaptic activity by forming voltage-dependent ion-selective channels [200–205]. Pardaxins are proposed to follow the barrel-stave model, with an aggregation number of 6 [198]. To test this hypothesis, we studied pardaxin 4 (Pa4), both in micelles and lipid bilayers, using a combination of solution and solid-state NMR spectroscopy [147]. Fast tumbling in the DPC micelle/Pa4 complex enabled us to obtain high-quality solution NMR data and to assign the majority of resonances for both the backbone and the side chains. We found that Pa4 adopts a bend-helix-bend-helix secondary structure motif, with an angle of ~122° between the two helical domains. The topological orientation of Pa4 in micelles was assessed by paramagnetic quenching experiments, establishing that Pa4 lies on the surface of the micelle.

This model is supported by solid-state NMR data in lipid membranes. Specifically, [13]C–[15]N rotational echo double resonance experiments (REDOR) carried out in vesicles of different composition (DMPC, POPC, and POPE:POPG 3:1 mixture) containing 3% Pa4 are consistent with peptide helical structure when embedded in membranes [206, 207]. ^{2}H and ^{31}P NMR experiments performed on POPC-d_{31} MLVs in the presence and absence of Pa4 revealed that upon peptide binding, disorder was increased both in headgroups and in acyl chains located in the hydrophobic core of the lipid membrane [197]. Interestingly, in DMPC, the C-terminal helix of Pa4 adopts a transmembrane orientation, while in POPC this domain is oriented with the helical axis approximately parallel to the bilayer normal. The latter suggests that membrane composition plays a pivotal role in the mechanism of action of Pa4. Taken together, the data from both solution and solid-state NMR corroborate the hypothesized "barrel-stave" mechanism of action of pardaxin, with the N-terminal domain involved in insertion into the bilayer, and the C-terminal helical portion involved in putative ion-channel formation.

4.3 Distinctin

This 47-residue peptide extracted from *Phyllomedusa distincta*, a tree frog from the Brazilian forests, interacts with negatively charged membranes and is active against both Gram-positive and Gram-negative bacteria. The primary sequence of distinctin comprises two linear chains of 22 (chain 1) and 25 (chain 2) residues, linked by a disulfide bridge between Cys19 of chain 1 and Cys23 of chain 2 [208–210]. Unlike other AMPs, distinctin adopts a well-folded conformation in aqueous environment, with a noncovalent parallel four-helical bundle that confers upon this peptide stability against proteolysis [211]. Scaloni and co-workers also showed that distinctin forms voltage-dependent ion channels in POPC/POPE planar bilayers [211], which were modeled as pentameric pores using molecular dynamics calculations [212]. The mechanisms of pore formation and membrane permeabilization have also been investigated using electrochemical measurements with two different mercury-supported biomimetic membranes, i.e., a self-assembled monolayer (SAM) and a tethered bilayer lipid membrane (tBLM). In SAM, distinctin forms selective ion channels for Tl^+ and Cd^{2+} ions; while the formation of K^+ permeable ion channels in tBLM occurred only at nonphysiological potentials [209].

Using a combination of site-specific ^{15}N and ^2H NMR spectroscopy, Bechinger and co-workers showed that distinctin's helical domains orient approximately parallel to the surface of mechanically oriented POPC (1-palmitoyl-2-oleoyl-*sn*-glycero-3-phosphocholine) lipid bilayers. In a parallel study, we confirmed Bechinger's findings and showed that distinctin interacts more strongly with membranes containing charged lipid headgroups (PA, PS, and PG). We also discovered that both chains of distinctin adopt an approximately parallel orientation with respect to the membrane plane, with a slight angle between chains 1 and 2. However, in 1:1 POPC:DOPA lipid bilayers and 50:1 lipid to protein molar ratio, distinctin is unable to disrupt lipid bilayers. The chemical shift anisotropy and dipolar couplings from separated local field experiments were implemented in a hybrid simulated annealing protocol [213] to determine the orientation of the distinctin heterodimer more quantitatively. After energy minimization, the tilt angles with respect to the bilayer plane were calculated to be ~24° and ~5° for chains 1 and 2, respectively [129].

Our findings in mechanically oriented bilayers are supported by experiments carried out on distinctin reconstituted in DMPC/DHPC bicelles. Under these experimental conditions, distinctin adopts a dual topology for chain 2, indicating that the peptide can be either absorbed on the surface of the bilayer or exist in a trans-membrane topology (Verardi et al. in preparation). These new findings support the electrophysiological data and explain how distinctin forms ion-conducting pores. In all, the data suggest that cell disruption may occur via formation in solution of stable tetramers that dissociate into monomers upon membrane interaction. Following an increase of concentration of the peptide on the membrane surface, distinctin organizes into ion-conducting pores. The propensity of distinctin to assume a transmembrane orientation even at low concentrations may suggest that stochastic formation of transient pores is also possible.

4.4   Cathelicidins

This well-known family of structurally different AMPs [37, 44, 214, 215] includes peptides that comprise an N-terminal signal peptide that shares a highly conserved cathelin-like domain and a variable cationic C-terminal domain, which is responsible for the antimicrobial activity. LL-37 is the only cathelicidin-derived peptide found in humans (LL-37-hCAP18) [44, 216–218]. LL-37-hCAP18 is expressed in epithelia, monocytes, and lymphocytes. During infection, inflammatory processes, or wound healing, a 100-residue pro-peptide is expressed, containing the signal peptide, the cathelin domains, and the 37-residue mature AMP located at the C-terminal region [37, 214, 217, 219]. LL-37 has a broad spectrum of bactericidal activity, may play a role in cystic fibrosis remediation, and has been found to inhibit HIV-1 infection in vitro [37, 220, 221].

Using DPC micelles, we were able to obtain high-resolution spectra of the mature LL-37 peptide [38]. The overall structure resembles that of Pa4, with a helix-break-helix motif and an angle between the two helical domains of ~120°. The N-terminal helix ends with a break at K12 and is more dynamic than the C-terminal helix (residues 13–33) and ends with a break at K12. The kink starting at residue 12 may be due to a hydrophobic cluster of residues located in the concave face of the peptide (I13, F17, and I20) and appears to be facilitated by a groove created by G14. Dynamic light scattering measurements show that addition of LL-37 to DPC micelles does not change the overall organization of micelles, which display an average hydrodynamic radius of 24 ± 2 Å and 26 ± 2 Å in the absence and presence of LL-37, respectively. The latter result suggests that LL-37 associates on the micelle surface without changing the overall micellar shape. These data were supported by paramagnetic quenching experiments carried out with Gd^{3+} and by the detection of several peptide-to-micelle NOEs. The overall topology for LL-37 is common to other natural occurring helical AMPs, with the hydrophilic residues pointing toward the bulk solvent and the hydrophobic toward the inner hydrocarbon core of the micelles. The structural topology of LL-37 in micelle is reported in Fig. 2. The peptide concave face containing the

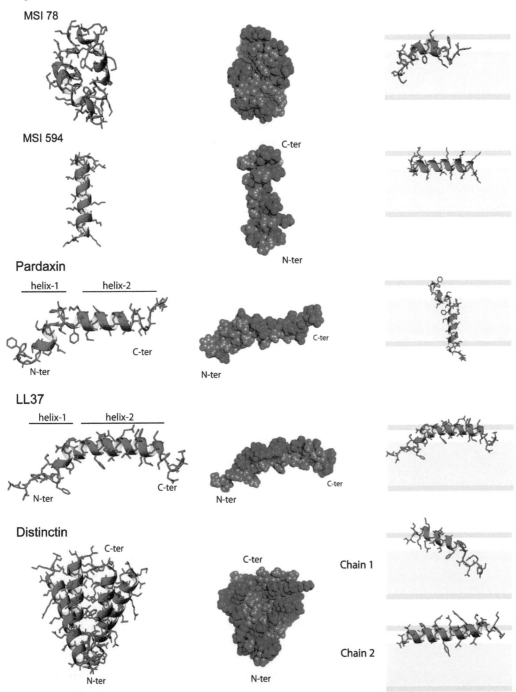

Fig. 2 Structures and membrane orientations (topology) of the helical AMPs studied by our groups. Note that the structures of MSI-78, MSI-594, Pa4, and LL-37 were obtained in detergent micelles, while the oligomeric structure of distinctin was obtained by in aqueous buffer. *Left*: backbone and side chain average structures of the AMPs from the NMR structural ensembles. *Center*: space-filled model of the structures with the hydrophilic residues colored *purple* and the hydrophobic in *blue*. *Right*: topology of the AMPs deduced from ssNMR experiments carried out in lipid membranes

hydrophobic cluster points toward the micelle interior with the hydrophilic residues pointing outward. The helix-kink-helix motif recurs in other AMPs (i.e., pardaxin), as already discussed. It has been hypothesized that the hinge region is required in order to confer structural flexibility for membrane insertion and pore formation [200, 202, 222, 223]. The large curvature present in LL-37 has also been observed in the magainin-derived peptides (MSI-78 and MSI-594) as well as in other lysine-rich peptides [57, 95, 224]. On the other hand, this curvature may be artificial and due to the interaction between the peptides and curved surfaces of the micelles. As described earlier in this article, micellar systems represent only a coarse approximation of membrane bilayers. Synthetic lipid bilayers such as vesicles or planar bilayers are preferable for testing lipid/peptide interactions. Despite structural similarities between LL-37 and pardaxins, their mechanisms of action appear to be different. Specifically, pardaxins are thought to disrupt cell membranes via a "barrel stave" mechanism, while LL-37 operates according to a "carpet-like" mechanism. The latter conclusion is supported by our helical model, as well as solid-state NMR studies from Ramamoorthy's laboratory [197, 225, 226]. Therefore, structure is important to modulate function, but amino acid composition is an important component that helps dictate the mechanism of action.

5   Conclusions and Perspectives

AMPs are emerging therapeutics with considerable potential. However, progress in the rational design of AMPs as powerful and selective drugs has been slow. A significant problem continues to be how to define the structural determinants for activities and specificities of action. Until this is understood, identification of the sequence features important for antimicrobial activity cannot be done on anything other than a trial-and-error basis. NMR is clearly playing a fundamental role in understanding the role of structure for AMPs, as reported in the recent literature both from our laboratory [129, 209] and others [52, 227–230]. The essential interactions of AMPs with membranes preclude the use of X-ray crystallography to determine high-resolution structural information. However, structure does not seem to be the only determining factor that accounts for the biological functions of AMPs. Primary sequence (amino acid content and nature), interactions with lipids, and flexibility are important factors to define specificity. Perhaps one of the most neglected aspects of research on AMPs is the characterization of their structural flexibilities. This is an area for which NMR is expected to have a special niche, pending the development of robust recombinant methods that will offer affordable combinations of isotope-labeled materials for solution and solid-state NMR.

Our working hypothesis is that AMPs are metamorphic polypeptides, capable of adopting different shapes. Thus, special structure may not be important per se. Instead, we suggest that the biological activities of AMPs may be encoded in their structural flexibility and ability to adopt several conformations and topologies upon homotropic (peptide–peptide) or heterotropic (peptide–lipid) interactions.

Acknowledgement

This work is partially supported by the National Institute of Health (GM 64742 to G.V.).

References

1. Bommarius B, Kalman D (2009) Antimicrobial and host defense peptides for therapeutic use against multidrug-resistant pathogens: new hope on the horizon. IDrugs 12:376–380

2. Bragonzi A (2010) Fighting back: peptidomimetics as a new weapon in the battle against antibiotic resistance. Sci Transl Med 2:21ps29

3. Giamarellou H, Poulakou G (2009) Multidrug-resistant Gram-negative infections: what are the treatment options? Drugs 69:1879–1901

4. Gryllos I, Tran-Winkler HJ, Cheng MF, Chung H, Bolcome R 3rd, Lu W, Lehrer RI, Wessels MR (2008) Induction of group A Streptococcus virulence by a human antimicrobial peptide. Proc Natl Acad Sci USA 105:16755–16760

5. Ho J, Tambyah PA, Paterson DL (2010) Multiresistant Gram-negative infections: a global perspective. Curr Opin Infect Dis 23:546–553

6. Paterson DL (2006) Clinical experience with recently approved antibiotics. Curr Opin Pharmacol 6:486–490

7. Cornaglia G, Rossolini GM (2009) Forthcoming therapeutic perspectives for infections due to multidrug-resistant Gram-positive pathogens. Clin Microbiol Infect 15:218–223

8. Wu G, Li X, Fan X, Wu H, Wang S, Shen Z, Xi T (2011) The activity of antimicrobial peptide S-thanatin is independent on multidrug-resistant spectrum of bacteria. Peptides 32:1139–1145

9. Marr AK, Gooderham WJ, Hancock RE (2006) Antibacterial peptides for therapeutic use: obstacles and realistic outlook. Curr Opin Pharmacol 6:468–472

10. Steiner H, Hultmark D, Engstrom A, Bennich H, Boman HG (1981) Sequence and specificity of two antibacterial proteins involved in insect immunity. Nature 292:246–248

11. Wiesner J, Vilcinskas A (2010) Antimicrobial peptides: the ancient arm of the human immune system. Virulence 1:440–464

12. Zasloff M (2007) Antimicrobial peptides, innate immunity, and the normally sterile urinary tract. J Am Soc Nephrol 18:2810–2816

13. Jenssen H, Hamill P, Hancock RE (2006) Peptide antimicrobial agents. Clin Microbiol Rev 19:491–511

14. Berridge NJ (1949) Preparation of the antibiotic nisin. Biochem J 45:486–493

15. Mattick AT, Hirsch A (1947) Further observations on an inhibitory substance (nisin) from lactic streptococci. Lancet 2:5–8

16. Ambatipudi K, Joss J, Raftery M, Deane E (2008) A proteomic approach to analysis of antimicrobial activity in marsupial pouch secretions. Dev Comp Immunol 32:108–120

17. Schittek B, Paulmann M, Senyurek I, Steffen H (2008) The role of antimicrobial peptides in human skin and in skin infectious diseases. Infect Disord Drug Targets 8:135–143

18. Steffen H, Rieg S, Wiedemann I, Kalbacher H, Deeg M, Sahl HG, Peschel A, Gotz F, Garbe C, Schittek B (2006) Naturally processed dermcidin-derived peptides do not permeabilize bacterial membranes and kill microorganisms irrespective of their charge. Antimicrob Agents Chemother 50:2608–2620

19. Hata TR, Gallo RL (2008) Antimicrobial peptides, skin infections, and atopic dermatitis. Semin Cutan Med Surg 27:144–150

20. Lai Y, Gallo RL (2009) AMPed up immunity: how antimicrobial peptides have multiple

roles in immune defense. Trends Immunol 30:131–141

21. Menousek J, Mishra B, Hanke ML, Heim CE, Kielian T, Wang G (2012) Database screening and in vivo efficacy of antimicrobial peptides against methicillin-resistant Staphylococcus aureus USA300. Int J Antimicrob Agents 39:402–406

22. Wang G, Li X, Wang Z (2009) APD2: the updated antimicrobial peptide database and its application in peptide design. Nucleic Acids Res 37:D933–D937

23. Abbanat D, Morrow B, Bush K (2008) New agents in development for the treatment of bacterial infections. Curr Opin Pharmacol 8:582–592

24. Hancock RE (1997) Peptide antibiotics. Lancet 349:418–422

25. Diamond G, Beckloff N, Weinberg A, Kisich KO (2009) The roles of antimicrobial peptides in innate host defense. Curr Pharm Des 15:2377–2392

26. Hancock RE, Diamond G (2000) The role of cationic antimicrobial peptides in innate host defences. Trends Microbiol 8:402–410

27. Barak O, Treat JR, James WD (2005) Antimicrobial peptides: effectors of innate immunity in the skin. Adv Dermatol 21:357–374

28. Izadpanah A, Gallo RL (2005) Antimicrobial peptides. J Am Acad Dermatol 52:381–390, quiz 391–382

29. Yin M, Gentili C, Koyama E, Zasloff M, Pacifici M (2002) Antiangiogenic treatment delays chondrocyte maturation and bone formation during limb skeletogenesis. J Bone Miner Res 17:56–65

30. Nissen-Meyer J, Nes IF (1997) Ribosomally synthesized antimicrobial peptides: their function, structure, biogenesis, and mechanism of action. Arch Microbiol 167:67–77

31. Papagianni M (2003) Ribosomally synthesized peptides with antimicrobial properties: biosynthesis, structure, function, and applications. Biotechnol Adv 21:465–499

32. Berkowitz BA, Bevins CL, Zasloff MA (1990) Magainins: a new family of membrane-active host defense peptides. Biochem Pharmacol 39:625–629

33. Boggs JM, Jo E, Polozov IV, Epand RF, Anantharamaiah GM, Blazyk J, Epand RM (2001) Effect of magainin, class L, and class A amphipathic peptides on fatty acid spin labels in lipid bilayers. Biochim Biophys Acta 1511:28–41

34. Porcelli F, Buck-Koehntop BA, Thennarasu S, Ramamoorthy A, Veglia G (2006) Structures of the dimeric and monomeric variants of magainin antimicrobial peptides (MSI-78 and MSI-594) in micelles and bilayers, determined by NMR spectroscopy. Biochemistry 45:5793–5799

35. Ramamoorthy A, Thennarasu S, Lee DK, Tan A, Maloy L (2006) Solid-state NMR investigation of the membrane-disrupting mechanism of antimicrobial peptides MSI-78 and MSI-594 derived from magainin 2 and melittin. Biophys J 91:206–216

36. Zasloff M (1987) Magainins, a class of antimicrobial peptides from Xenopus skin: isolation, characterization of two active forms, and partial cDNA sequence of a precursor. Proc Natl Acad Sci USA 84:5449–5453

37. Nijnik A, Hancock RE (2009) The roles of cathelicidin LL-37 in immune defences and novel clinical applications. Curr Opin Hematol 16:41–47

38. Porcelli F, Verardi R, Shi L, Henzler-Wildman KA, Ramamoorthy A, Veglia G (2008) NMR structure of the cathelicidin-derived human antimicrobial peptide LL-37 in dodecylphosphocholine micelles. Biochemistry 47:5565–5572

39. Bals R, Wang X, Wu Z, Freeman T, Bafna V, Zasloff M, Wilson JM (1998) Human beta-defensin 2 is a salt-sensitive peptide antibiotic expressed in human lung. J Clin Invest 102:874–880

40. Howell MD (2007) The role of human beta defensins and cathelicidins in atopic dermatitis. Curr Opin Allergy Clin Immunol 7:413–417

41. Tang M, Waring AJ, Lehrer RI, Hong M (2006) Orientation of a beta-hairpin antimicrobial peptide in lipid bilayers from two-dimensional dipolar chemical-shift correlation NMR. Biophys J 90:3616–3624

42. Steinstraesser L, Kraneburg UM, Hirsch T, Kesting M, Steinau HU, Jacobsen F, Al-Benna S (2009) Host defense peptides as effector molecules of the innate immune response: a sledgehammer for drug resistance? Int J Mol Sci 10:3951–3970

43. Boman HG (1995) Peptide antibiotics and their role in innate immunity. Annu Rev Immunol 13:61–92

44. Guani-Guerra E, Santos-Mendoza T, Lugo-Reyes SO, Teran LM (2010) Antimicrobial peptides: general overview and clinical implications in human health and disease. Clin Immunol 135:1–11

45. Buffy JJ, Hong T, Yamaguchi S, Waring AJ, Lehrer RI, Hong M (2003) Solid-state NMR investigation of the depth of insertion of

protegrin-1 in lipid bilayers using paramagnetic Mn2+. Biophys J 85:2363–2373

46. Chekmenev EY, Jones SM, Nikolayeva YN, Vollmar BS, Wagner TJ, Gor'kov PL, Brey WW, Manion MN, Daugherty KC, Cotten M (2006) High-field NMR studies of molecular recognition and structure-function relationships in antimicrobial piscidins at the water-lipid bilayer interface. J Am Chem Soc 128:5308–5309

47. Bonev BB, Chan WC, Bycroft BW, Roberts GC, Watts A (2000) Interaction of the lantibiotic nisin with mixed lipid bilayers: a 31P and 2H NMR study. Biochemistry 39:11425–11433

48. Almeida PF, Pokorny A (2009) Mechanisms of antimicrobial, cytolytic, and cell-penetrating peptides: from kinetics to thermodynamics. Biochemistry 48:8083–8093

49. Bechinger B (2011) Insights into the mechanisms of action of host defence peptides from biophysical and structural investigations. J Pept Sci 17:306–314

50. Brogden KA (2005) Antimicrobial peptides: pore formers or metabolic inhibitors in bacteria? Nat Rev Microbiol 3:238–250

51. Dathe M, Wieprecht T (1999) Structural features of helical antimicrobial peptides: their potential to modulate activity on model membranes and biological cells. Biochim Biophys Acta 1462:71–87

52. Bechinger B, Salnikov ES (2012) The membrane interactions of antimicrobial peptides revealed by solid-state NMR spectroscopy. Chem Phys Lipids 165:282–301

53. Blazyk J, Wiegand R, Klein J, Hammer J, Epand RM, Epand RF, Maloy WL, Kari UP (2001) A novel linear amphipathic beta-sheet cationic antimicrobial peptide with enhanced selectivity for bacterial lipids. J Biol Chem 276:27899–27906

54. Epand RF, Epand RM, Monaco V, Stoia S, Formaggio F, Crisma M, Toniolo C (1999) The antimicrobial peptide trichogin and its interaction with phospholipid membranes. Eur J Biochem 266:1021–1028

55. Boland MP, Separovic F (2006) Membrane interactions of antimicrobial peptides from Australian tree frogs. Biochim Biophys Acta 1758:1178–1183

56. Epand RF, Umezawa N, Porter EA, Gellman SH, Epand RM (2003) Interactions of the antimicrobial beta-peptide beta-17 with phospholipid vesicles differ from membrane interactions of magainins. Eur J Biochem 270:1240–1248

57. Fernandez DI, Sani MA, Gehman JD, Hahm KS, Separovic F (2011) Interactions of a synthetic Leu-Lys-rich antimicrobial peptide with phospholipid bilayers. Eur Biophys J 40:471–480

58. Gazit E, Boman A, Boman HG, Shai Y (1995) Interaction of the mammalian antibacterial peptide cecropin P1 with phospholipid vesicles. Biochemistry 34:11479–11488

59. Jiang Z, Kullberg BJ, van der Lee H, Vasil AI, Hale JD, Mant CT, Hancock RE, Vasil ML, Netea MG, Hodges RS (2008) Effects of hydrophobicity on the antifungal activity of alpha-helical antimicrobial peptides. Chem Biol Drug Des 72:483–495

60. Dathe M, Wieprecht T, Nikolenko H, Handel L, Maloy WL, MacDonald DL, Beyermann M, Bienert M (1997) Hydrophobicity, hydrophobic moment and angle subtended by charged residues modulate antibacterial and haemolytic activity of amphipathic helical peptides. FEBS Lett 403:208–212

61. Hancock RE, Rozek A (2002) Role of membranes in the activities of antimicrobial cationic peptides. FEMS Microbiol Lett 206:143–149

62. Almeida PF, Pokorny A (2010) Binding and permeabilization of model membranes by amphipathic peptides. Methods Mol Biol 618:155–169

63. Epand RF, Pollard JE, Wright JO, Savage PB, Epand RM (2010) Depolarization, bacterial membrane composition, and the antimicrobial action of ceragenins. Antimicrob Agents Chemother 54:3708–3713

64. Epand RF, Schmitt MA, Gellman SH, Epand RM (2006) Role of membrane lipids in the mechanism of bacterial species selective toxicity by two alpha/beta-antimicrobial peptides. Biochim Biophys Acta 1758:1343–1350

65. Shai Y (2002) Mode of action of membrane active antimicrobial peptides. Biopolymers 66:236–248

66. Shai Y (1999) Mechanism of the binding, insertion and destabilization of phospholipid bilayer membranes by alpha-helical antimicrobial and cell non-selective membrane-lytic peptides. Biochim Biophys Acta 1462:55–70

67. Papo N, Shai Y (2003) Can we predict biological activity of antimicrobial peptides from their interactions with model phospholipid membranes? Peptides 24:1693–1703

68. Teixeira V, Feio MJ, Bastos M (2012) Role of lipids in the interaction of antimicrobial peptides with membranes. Prog Lipid Res 51:149–177

69. Epand RM, Epand RF (2011) Bacterial membrane lipids in the action of antimicrobial agents. J Pept Sci 17:298–305

70. Aisenbrey C, Bertani P, Bechinger B (2010) Solid-state NMR investigations of membrane-associated antimicrobial peptides. Methods Mol Biol 618:209–233

71. Bechinger B (2010) Membrane association and pore formation by alpha-helical peptides. Adv Exp Med Biol 677:24–30

72. Huang HW (2006) Molecular mechanism of antimicrobial peptides: the origin of cooperativity. Biochim Biophys Acta 1758:1292–1302

73. Huang HW (2000) Action of antimicrobial peptides: two-state model. Biochemistry 39:8347–8352

74. Bechinger B (2000) Biophysical investigations of membrane perturbations by polypeptides using solid-state NMR spectroscopy (review). Mol Membr Biol 17:135–142

75. Chen FY, Lee MT, Huang HW (2003) Evidence for membrane thinning effect as the mechanism for peptide-induced pore formation. Biophys J 84:3751–3758

76. Pabst G, Grage SL, Danner-Pongratz S, Jing W, Ulrich AS, Watts A, Lohner K, Hickel A (2008) Membrane thickening by the antimicrobial peptide PGLa. Biophys J 95:5779–5788

77. Nomura K, Corzo G (2006) The effect of binding of spider-derived antimicrobial peptides, oxyopinins, on lipid membranes. Biochim Biophys Acta 1758:1475–1482

78. Nomura K, Corzo G, Nakajima T, Iwashita T (2004) Orientation and pore-forming mechanism of a scorpion pore-forming peptide bound to magnetically oriented lipid bilayers. Biophys J 87:2497–2507

79. Smith PE, Brender JR, Ramamoorthy A (2009) Induction of negative curvature as a mechanism of cell toxicity by amyloidogenic peptides: the case of islet amyloid polypeptide. J Am Chem Soc 131:4470–4478

80. Wieprecht T, Beyermann M, Seelig J (2002) Thermodynamics of the coil-alpha-helix transition of amphipathic peptides in a membrane environment: the role of vesicle curvature. Biophys Chem 96:191–201

81. Wi S, Kim C (2008) Pore structure, thinning effect, and lateral diffusive dynamics of oriented lipid membranes interacting with antimicrobial peptide protegrin-1: 31P and 2H solid-state NMR study. J Phys Chem B 112:11402–11414

82. White SH, Wimley WC (1999) Membrane protein folding and stability: physical principles. Annu Rev Biophys Biomol Struct 28:319–365

83. Seelig J (2004) Thermodynamics of lipid-peptide interactions. Biochim Biophys Acta 1666:40–50

84. Wimley WC (2010) Describing the mechanism of antimicrobial peptide action with the interfacial activity model. ACS Chem Biol 5:905–917

85. Hristova K, Wimley WC (2011) A look at arginine in membranes. J Membr Biol 239:49–56

86. Oren Z, Shai Y (1998) Mode of action of linear amphipathic alpha-helical antimicrobial peptides. Biopolymers 47:451–463

87. Shai Y, Oren Z (2001) From "carpet" mechanism to de-novo designed diastereomeric cell-selective antimicrobial peptides. Peptides 22:1629–1641

88. Avrahami D, Oren Z, Shai Y (2001) Effect of multiple aliphatic amino acids substitutions on the structure, function, and mode of action of diastereomeric membrane active peptides. Biochemistry 40:12591–12603

89. Pokorny A, Almeida PF (2004) Kinetics of dye efflux and lipid flip-flop induced by delta-lysin in phosphatidylcholine vesicles and the mechanism of graded release by amphipathic, alpha-helical peptides. Biochemistry 43:8846–8857

90. Oren Z, Lerman JC, Gudmundsson GH, Agerberth B, Shai Y (1999) Structure and organization of the human antimicrobial peptide LL-37 in phospholipid membranes: relevance to the molecular basis for its non-cell-selective activity. Biochem J 341 (Pt 3):501–513

91. He K, Ludtke SJ, Worcester DL, Huang HW (1996) Neutron scattering in the plane of membranes: structure of alamethicin pores. Biophys J 70:2659–2666

92. Gehman JD, Luc F, Hall K, Lee TH, Boland MP, Pukala TL, Bowie JH, Aguilar MI, Separovic F (2008) Effect of antimicrobial peptides from Australian tree frogs on anionic phospholipid membranes. Biochemistry 47:8557–8565

93. Ludtke SJ, He K, Heller WT, Harroun TA, Yang L, Huang HW (1996) Membrane pores induced by magainin. Biochemistry 35:13723–13728

94. Yang L, Harroun TA, Weiss TM, Ding L, Huang HW (2001) Barrel-stave model or toroidal model? A case study on melittin pores. Biophys J 81:1475–1485

95. Epand RF, Maloy L, Ramamoorthy A, Epand RM (2010) Amphipathic helical cationic antimicrobial peptides promote rapid formation

of crystalline states in the presence of phosphatidylglycerol: lipid clustering in anionic membranes. Biophys J 98:2564–2573

96. Epand RF, Maloy WL, Ramamoorthy A, Epand RM (2010) Probing the "charge cluster mechanism" in amphipathic helical cationic antimicrobial peptides. Biochemistry 49:4076–4084

97. Pag U, Oedenkoven M, Sass V, Shai Y, Shamova O, Antcheva N, Tossi A, Sahl HG (2008) Analysis of in vitro activities and modes of action of synthetic antimicrobial peptides derived from an alpha-helical 'sequence template'. J Antimicrob Chemother 61:341–352

98. Wieprecht T, Apostolov O, Beyermann M, Seelig J (2000) Membrane binding and pore formation of the antibacterial peptide PGLa: thermodynamic and mechanistic aspects. Biochemistry 39:442–452

99. Yandek LE, Pokorny A, Floren A, Knoelke K, Langel U, Almeida PF (2007) Mechanism of the cell-penetrating peptide transportan 10 permeation of lipid bilayers. Biophys J 92:2434–2444

100. Ladokhin AS, Wimley WC, Hristova K, White SH (1997) Mechanism of leakage of contents of membrane vesicles determined by fluorescence requenching. Methods Enzymol 278:474–486

101. Ladokhin AS, Wimley WC, White SH (1995) Leakage of membrane vesicle contents: determination of mechanism using fluorescence requenching. Biophys J 69:1964–1971

102. Gregory SM, Cavenaugh A, Journigan V, Pokorny A, Almeida PF (2008) A quantitative model for the all-or-none permeabilization of phospholipid vesicles by the antimicrobial peptide cecropin A. Biophys J 94:1667–1680

103. Gregory SM, Pokorny A, Almeida PF (2009) Magainin 2 revisited: a test of the quantitative model for the all-or-none permeabilization of phospholipid vesicles. Biophys J 96:116–131

104. Wimley WC, Selsted ME, White SH (1994) Interactions between human defensins and lipid bilayers: evidence for formation of multimeric pores. Protein Sci 3:1362–1373

105. Epand RM, Vogel HJ (1999) Diversity of antimicrobial peptides and their mechanisms of action. Biochim Biophys Acta 1462:11–28

106. Koprivnjak T, Peschel A (2011) Bacterial resistance mechanisms against host defense peptides. Cell Mol Life Sci 68:2243–2254

107. Marion D, Zasloff M, Bax A (1988) A two-dimensional NMR study of the antimicrobial peptide magainin 2. FEBS Lett 227:21–26

108. Bechinger B (1999) The structure, dynamics and orientation of antimicrobial peptides in membranes by multidimensional solid-state NMR spectroscopy. Biochim Biophys Acta 1462:157–183

109. Hauge HH, Mantzilas D, Eijsink VG, Nissen-Meyer J (1999) Membrane-mimicking entities induce structuring of the two-peptide bacteriocins plantaricin E/F and plantaricin J/K. J Bacteriol 181:740–747

110. Luo P, Baldwin RL (1997) Mechanism of helix induction by trifluoroethanol: a framework for extrapolating the helix-forming properties of peptides from trifluoroethanol/water mixtures back to water. Biochemistry 36:8413–8421

111. Prosser RS, Evanics F, Kitevski JL, Al-Abdul-Wahid MS (2006) Current applications of bicelles in NMR studies of membrane-associated amphiphiles and proteins. Biochemistry 45:8453–8465

112. Warschawski DE, Arnold AA, Beaugrand M, Gravel A, Chartrand E, Marcotte I (2011) Choosing membrane mimetics for NMR structural studies of transmembrane proteins. Biochim Biophys Acta 1808:1957–1974

113. Bourbigot S, Fardy L, Waring AJ, Yeaman MR, Booth V (2009) Structure of chemokine-derived antimicrobial peptide interleukin-8alpha and interaction with detergent micelles and oriented lipid bilayers. Biochemistry 48:10509–10521

114. Khandelia H, Kaznessis YN (2005) Molecular dynamics simulations of helical antimicrobial peptides in SDS micelles: what do point mutations achieve? Peptides 26:2037–2049

115. Mascioni A, Porcelli F, Ilangovan U, Ramamoorthy A, Veglia G (2003) Conformational preferences of the amylin nucleation site in SDS micelles: an NMR study. Biopolymers 69:29–41

116. Buffy JJ, Buck-Koehntop BA, Porcelli F, Traaseth NJ, Thomas DD, Veglia G (2006) Defining the intramembrane binding mechanism of sarcolipin to calcium ATPase using solution NMR spectroscopy. J Mol Biol 358:420–429

117. Sherman PJ, Jackway RJ, Gehman JD, Praporski S, McCubbin GA, Mechler A, Martin LL, Separovic F, Bowie JH (2009) Solution structure and membrane interactions of the antimicrobial peptide fallaxidin 4.1a: an NMR and QCM study. Biochemistry 48:11892–11901

118. Poget SF, Cahill SM, Girvin ME (2007) Isotropic bicelles stabilize the functional form of a small multidrug-resistance pump for NMR structural studies. J Am Chem Soc 129:2432–2433

119. Sanders CR 2nd, Landis GC (1995) Reconstitution of membrane proteins into lipid-rich bilayered mixed micelles for NMR studies. Biochemistry 34:4030–4040

120. Sanders CR, Prosser RS (1998) Bicelles: a model membrane system for all seasons? Structure 6:1227–1234

121. Marcotte I, Wegener KL, Lam YH, Chia BC, de Planque MR, Bowie JH, Auger M, Separovic F (2003) Interaction of antimicrobial peptides from Australian amphibians with lipid membranes. Chem Phys Lipids 122: 107–120

122. Dittmer J, Thogersen L, Underhaug J, Bertelsen K, Vosegaard T, Pedersen JM, Schiott B, Tajkhorshid E, Skrydstrup T, Nielsen NC (2009) Incorporation of antimicrobial peptides into membranes: a combined liquid-state NMR and molecular dynamics study of alamethicin in DMPC/DHPC bicelles. J Phys Chem B 113:6928–6937

123. Jiang Y, Wang H, Kindt JT (2010) Atomistic simulations of bicelle mixtures. Biophys J 98:2895–2903

124. van Dam L, Karlsson G, Edwards K (2006) Morphology of magnetically aligning DMPC/DHPC aggregates-perforated sheets, not disks. Langmuir 22:3280–3285

125. Bechinger B (2005) Detergent-like properties of magainin antibiotic peptides: a 31P solid-state NMR spectroscopy study. Biochim Biophys Acta 1712:101–108

126. Cardon TB, Tiburu EK, Lorigan GA (2003) Magnetically aligned phospholipid bilayers in weak magnetic fields: optimization, mechanism, and advantages for X-band EPR studies. J Magn Reson 161:77–90

127. Diller A, Loudet C, Aussenac F, Raffard G, Fournier S, Laguerre M, Grelard A, Opella SJ, Marassi FM, Dufourc EJ (2009) Bicelles: a natural 'molecular goniometer' for structural, dynamical and topological studies of molecules in membranes. Biochimie 91:744–751

128. De Angelis AA, Grant CV, Baxter MK, McGavin JA, Opella SJ, Cotten ML (2011) Amphipathic antimicrobial piscidin in magnetically aligned lipid bilayers. Biophys J 101:1086–1094

129. Verardi R, Traaseth NJ, Shi L, Porcelli F, Monfregola L, De Luca S, Amodeo P, Veglia G, Scaloni A (2012) Probing membrane topology of the antimicrobial peptide distinctin by solid-state NMR spectroscopy in zwitterionic and charged lipid bilayers. Biochim Biophys Acta 1808:34–40

130. Mote KR, Gopinath T, Traaseth NJ, Kitchen J, Gor'kov PL, Brey WW, Veglia G (2011) Multidimensional oriented solid-state NMR experiments enable the sequential assignment of uniformly 15N labeled integral membrane proteins in magnetically aligned lipid bilayers. J Biomol NMR 51:339–346

131. Bechinger B, Kim Y, Chirlian LE, Gesell J, Neumann JM, Montal M, Tomich J, Zasloff M, Opella SJ (1991) Orientations of amphipathic helical peptides in membrane bilayers determined by solid-state NMR spectroscopy. J Biomol NMR 1:167–173

132. Opella SJ, Ma C, Marassi FM (2001) Nuclear magnetic resonance of membrane-associated peptides and proteins. Methods Enzymol 339:285–313

133. Valentine KG, Liu SF, Marassi FM, Veglia G, Opella SJ, Ding FX, Wang SH, Arshava B, Becker JM, Naider F (2001) Structure and topology of a peptide segment of the 6th transmembrane domain of the Saccharomyces cerevisae alpha-factor receptor in phospholipid bilayers. Biopolymers 59:243–256

134. Killian JA, Borle F, de Kruijff B, Seelig J (1986) Comparative 2H- and 31P-NMR study on the properties of palmitoyllysophosphatidylcholine in bilayers with gramicidin, cholesterol and dipalmitoylphosphatidylcholine. Biochim Biophys Acta 854:133–142

135. Klocek G, Schulthess T, Shai Y, Seelig J (2009) Thermodynamics of melittin binding to lipid bilayers. Aggregation and pore formation. Biochemistry 48:2586–2596

136. Seelig J, MacDonald PM (1987) Phospholipids and proteins in biological membranes. 2H NMR as a method to study structure, dynamics, and interactions. Acc Chem Res 20:221–228

137. Sharma M, Yi M, Dong H, Qin H, Peterson E, Busath DD, Zhou HX, Cross TA (2010) Insight into the mechanism of the influenza A proton channel from a structure in a lipid bilayer. Science 330:509–512

138. Mascioni A, Karim C, Barany G, Thomas DD, Veglia G (2002) Structure and orientation of sarcolipin in lipid environments. Biochemistry 41:475–482

139. Mascioni A, Karim C, Zamoon J, Thomas DD, Veglia G (2002) Solid-state NMR and rigid body molecular dynamics to determine domain orientations of monomeric phospholamban. J Am Chem Soc 124:9392–9393

140. Traaseth NJ, Buffy JJ, Zamoon J, Veglia G (2006) Structural dynamics and topology of phospholamban in oriented lipid bilayers using multidimensional solid-state NMR. Biochemistry 45:13827–13834

141. Traaseth NJ, Shi L, Verardi R, Mullen DG, Barany G, Veglia G (2009) Structure and

topology of monomeric phospholamban in lipid membranes determined by a hybrid solution and solid-state NMR approach. Proc Natl Acad Sci USA 106:10165–10170

142. Verardi R, Shi L, Traaseth NJ, Walsh N, Veglia G (2011) Structural topology of phospholamban pentamer in lipid bilayers by a hybrid solution and solid-state NMR method. Proc Natl Acad Sci USA 108:9101–9106

143. Epand RM, Epand RF (2003) Liposomes as models for antimicrobial peptides. Methods Enzymol 372:124–133

144. Marquette A, Lorber B, Bechinger B (2010) Reversible liposome association induced by LAH4: a peptide with potent antimicrobial and nucleic acid transfection activities. Biophys J 98:2544–2553

145. Georgescu J, Munhoz VH, Bechinger B (2010) NMR structures of the histidine-rich peptide LAH4 in micellar environments: membrane insertion, pH-dependent mode of antimicrobial action, and DNA transfection. Biophys J 99:2507–2515

146. Malliavin TE, Giudice E (2002) Analysis of peptide rotational diffusion by homonuclear NMR. Biopolymers 63:335–342

147. Porcelli F, Buck B, Lee DK, Hallock KJ, Ramamoorthy A, Veglia G (2004) Structure and orientation of pardaxin determined by NMR experiments in model membranes. J Biol Chem 279:45815–45823

148. Park SH, Kim YK, Park JW, Lee B, Lee BJ (2000) Solution structure of the antimicrobial peptide gaegurin 4 by H and 15N nuclear magnetic resonance spectroscopy. Eur J Biochem 267:2695–2704

149. Matsumori N, Murata M (2010) 3D structures of membrane-associated small molecules as determined in isotropic bicelles. Nat Prod Rep 27:1480–1492

150. Yamamoto K, Vivekanandan S, Ramamoorthy A (2011) Fast NMR data acquisition from bicelles containing a membrane-associated peptide at natural-abundance. J Phys Chem B 115:12448–12455

151. Chou JJ, Delaglio F, Bax A (2000) Measurement of one-bond 15N-13C' dipolar couplings in medium sized proteins. J Biomol NMR 18:101–105

152. Kubat JA, Chou JJ, Rovnyak D (2007) Nonuniform sampling and maximum entropy reconstruction applied to the accurate measurement of residual dipolar couplings. J Magn Reson 186:201–211

153. Jaroniec CP, Boisbouvier J, Tworowska I, Nikonowicz EP, Bax A (2005) Accurate measurement of 15N-13C residual dipolar couplings in nucleic acids. J Biomol NMR 31:231–241

154. Al-Abdul-Wahid MS, Verardi R, Veglia G, Prosser RS (2011) Topology and immersion depth of an integral membrane protein by paramagnetic rates from dissolved oxygen. J Biomol NMR 51:173–183

155. Donghi D, Sigel RK (2012) Metal ion-RNA interactions studied via multinuclear NMR. Methods Mol Biol 848:253–273

156. Glover KJ, Whiles JA, Vold RR, Melacini G (2002) Position of residues in transmembrane peptides with respect to the lipid bilayer: a combined lipid Noes and water chemical exchange approach in phospholipid bicelles. J Biomol NMR 22:57–64

157. Huang H, Melacini G (2006) High-resolution protein hydration NMR experiments: probing how protein surfaces interact with water and other non-covalent ligands. Anal Chim Acta 564:1–9

158. Melacini G, Boelens R, Kaptein R (1999) Band-selective editing of exchange-relay in protein-water NOE experiments. J Biomol NMR 13:67–71

159. Choutko A, Glattli A, Fernandez C, Hilty C, Wuthrich K, van Gunsteren WF (2010) Membrane protein dynamics in different environments: simulation study of the outer membrane protein X in a lipid bilayer and in a micelle. Eur Biophys J 40:39–58

160. Fernandez C, Adeishvili K, Wuthrich K (2001) Transverse relaxation-optimized NMR spectroscopy with the outer membrane protein OmpX in dihexanoyl phosphatidyl-choline micelles. Proc Natl Acad Sci USA 98:2358–2363

161. Fernandez C, Hilty C, Wider G, Wuthrich K (2002) Lipid-protein interactions in DHPC micelles containing the integral membrane protein OmpX investigated by NMR spectroscopy. Proc Natl Acad Sci USA 99:13533–13537

162. Fernandez C, Wuthrich K (2003) NMR solution structure determination of membrane proteins reconstituted in detergent micelles. FEBS Lett 555:144–150

163. Nishida N, Shimada I (2011) An NMR method to study protein-protein interactions. Methods Mol Biol 757:129–137

164. Shimada I (2005) NMR techniques for identifying the interface of a larger protein-protein complex: cross-saturation and transferred cross-saturation experiments. Methods Enzymol 394:483–506

165. Takahashi H, Miyazawa M, Ina Y, Fukunishi Y, Mizukoshi Y, Nakamura H, Shimada I

(2006) Utilization of methyl proton resonances in cross-saturation measurement for determining the interfaces of large protein-protein complexes. J Biomol NMR 34: 167–177

166. Cook GA, Zhang H, Park SH, Wang Y, Opella SJ (2010) Comparative NMR studies demonstrate profound differences between two viroporins: p7 of HCV and Vpu of HIV-1. Biochim Biophys Acta 1808:554–560

167. Howard KP, Opella SJ (1996) High-resolution solid-state NMR spectra of integral membrane proteins reconstituted into magnetically oriented phospholipid bilayers. J Magn Reson B 112:91–94

168. Gopinath T, Traaseth NJ, Mote K, Veglia G (2010) Sensitivity enhanced heteronuclear correlation spectroscopy in multidimensional solid-state NMR of oriented systems via chemical shift coherences. J Am Chem Soc 132:5357–5363

169. Gopinath T, Veglia G (2012) Dual acquisition magic-angle spinning solid-state NMR-spectroscopy: simultaneous acquisition of multidimensional spectra of biomacromolecules. Angew Chem Int Ed Engl 51: 2731–2735

170. Bechinger B, Gierasch LM, Montal M, Zasloff M, Opella SJ (1996) Orientations of helical peptides in membrane bilayers by solid state NMR spectroscopy. Solid State Nucl Magn Reson 7:185–191

171. Bechinger B, Skladnev DA, Ogrel A, Li X, Rogozhkina EV, Ovchinnikova TV, O'Neil JD, Raap J (2001) 15N and 31P solid-state NMR investigations on the orientation of zervamicin II and alamethicin in phosphatidylcholine membranes. Biochemistry 40: 9428–9437

172. Bechinger B, Zasloff M, Opella SJ (1993) Structure and orientation of the antibiotic peptide magainin in membranes by solid-state nuclear magnetic resonance spectroscopy. Protein Sci 2:2077–2084

173. Gustavsson M, Traaseth NJ, Veglia G (2012) Probing ground and excited states of phospholamban in model and native lipid membranes by magic angle spinning NMR spectroscopy. Biochim Biophys Acta 1818:146–153

174. Tang M, Hong M (2009) Structure and mechanism of beta-hairpin antimicrobial peptides in lipid bilayers from solid-state NMR spectroscopy. Mol Biosyst 5:317–322

175. Yamaguchi S, Huster D, Waring A, Lehrer RI, Kearney W, Tack BF, Hong M (2001) Orientation and dynamics of an antimicrobial peptide in the lipid bilayer by solid-state

NMR spectroscopy. Biophys J 81: 2203–2214

176. Ramamoorthy A, Lee DK, Santos JS, Henzler-Wildman KA (2008) Nitrogen-14 solid-state NMR spectroscopy of aligned phospholipid bilayers to probe peptide-lipid interaction and oligomerization of membrane associated peptides. J Am Chem Soc 130:11023–11029

177. Thennarasu S, Lee DK, Poon A, Kawulka KE, Vederas JC, Ramamoorthy A (2005) Membrane permeabilization, orientation, and antimicrobial mechanism of subtilosin A. Chem Phys Lipids 137:38–51

178. Ketchem RR, Hu W, Cross TA (1993) High-resolution conformation of gramicidin A in a lipid bilayer by solid-state NMR. Science 261:1457–1460

179. Quine JR, Brenneman MT, Cross TA (1997) Protein structural analysis from solid-state NMR-derived orientational constraints. Biophys J 72:2342–2348

180. Bechinger B, Kinder R, Helmle M, Vogt TC, Harzer U, Schinzel S (1999) Peptide structural analysis by solid-state NMR spectroscopy. Biopolymers 51:174–190

181. Marassi FM, Ma C, Gesell JJ, Opella SJ (2000) Three-dimensional solid-state NMR spectroscopy is essential for resolution of resonances from in-plane residues in uniformly (15)N-labeled helical membrane proteins in oriented lipid bilayers. J Magn Reson 144:156–161

182. De Angelis AA, Nevzorov AA, Park SH, Howell SC, Mrse AA, Opella SJ (2004) High-resolution NMR spectroscopy of membrane proteins in aligned bicelles. J Am Chem Soc 126:15340–15341

183. Andronesi OC, Pfeifer JR, Al-Momani L, Ozdirekcan S, Rijkers DT, Angerstein B, Luca S, Koert U, Killian JA, Baldus M (2004) Probing membrane protein orientation and structure using fast magic-angle-spinning solid-state NMR. J Biomol NMR 30: 253–265

184. Elena B, Hediger S, Emsley L (2003) Correlation of fast and slow chemical shift spinning sideband patterns under fast magic-angle spinning. J Magn Reson 160:40–46

185. Griffin RG (1998) Dipolar recoupling in MAS spectra of biological solids. Nat Struct Biol 5(Suppl):508–512

186. Zasloff M, Martin B, Chen HC (1988) Antimicrobial activity of synthetic magainin peptides and several analogues. Proc Natl Acad Sci USA 85:910–913

187. Bechinger B, Zasloff M, Opella SJ (1998) Structure and dynamics of the antibiotic

peptide PGLa in membranes by solution and solid-state nuclear magnetic resonance spectroscopy. Biophys J 74:981–987

188. Bechinger B, Zasloff M, Opella SJ (1992) Structure and interactions of magainin antibiotic peptides in lipid bilayers: a solid-state nuclear magnetic resonance investigation. Biophys J 62:12–14

189. Gesell J, Zasloff M, Opella SJ (1997) Two-dimensional 1H NMR experiments show that the 23-residue magainin antibiotic peptide is an alpha-helix in dodecylphosphocholine micelles, sodium dodecylsulfate micelles, and trifluoroethanol/water solution. J Biomol NMR 9:127–135

190. Matsuzaki K, Mitani Y, Akada KY, Murase O, Yoneyama S, Zasloff M, Miyajima K (1998) Mechanism of synergism between antimicrobial peptides magainin 2 and PGLa. Biochemistry 37:15144–15153

191. Ramamoorthy A, Marassi FM, Zasloff M, Opella SJ (1995) Three-dimensional solid-state NMR spectroscopy of a peptide oriented in membrane bilayers. J Biomol NMR 6:329–334

192. Wieprecht T, Apostolov O, Seelig J (2000) Binding of the antibacterial peptide magainin 2 amide to small and large unilamellar vesicles. Biophys Chem 85:187–198

193. Islam K, Hawser SP (1998) MSI-78 magainin pharmaceuticals. IDrugs 1:605–609

194. Yang P, Ramamoorthy A, Chen Z (2011) Membrane orientation of MSI-78 measured by sum frequency generation vibrational spectroscopy. Langmuir 27:7760–7767

195. Domadia PN, Bhunia A, Ramamoorthy A, Bhattacharjya S (2010) Structure, interactions, and antibacterial activities of MSI-594 derived mutant peptide MSI-594F5A in lipopolysaccharide micelles: role of the helical hairpin conformation in outer-membrane permea-bilization. J Am Chem Soc 132: 18417–18428

196. Wakamatsu K, Takeda A, Tachi T, Matsuzaki K (2002) Dimer structure of magainin 2 bound to phospholipid vesicles. Biopolymers 64:314–327

197. Henzler-Wildman KA, Martinez GV, Brown MF, Ramamoorthy A (2004) Perturbation of the hydrophobic core of lipid bilayers by the human antimicrobial peptide LL-37. Biochemistry 43:8459–8469

198. Lazarovici P, Primor N, Loew LM (1986) Purification and pore-forming activity of two hydrophobic polypeptides from the secretion of the Red Sea Moses sole (Pardachirus marmoratus). J Biol Chem 261: 16704–16713

199. Oren Z, Shai Y (1996) A class of highly potent antibacterial peptides derived from pardaxin, a pore-forming peptide isolated from Moses sole fish Pardachirus marmoratus. Eur J Biochem 237:303–310

200. Bhunia A, Domadia PN, Torres J, Hallock KJ, Ramamoorthy A, Bhattacharjya S (2010) NMR structure of pardaxin, a pore-forming antimicrobial peptide, in lipopolysaccharide micelles: mechanism of outer membrane permeabilization. J Biol Chem 285:3883–3895

201. Epand RF, Ramamoorthy A, Epand RM (2006) Membrane lipid composition and the interaction of pardaxin: the role of cholesterol. Protein Pept Lett 13:1–5

202. Hallock KJ, Lee DK, Omnaas J, Mosberg HI, Ramamoorthy A (2002) Membrane composition determines pardaxin's mechanism of lipid bilayer disruption. Biophys J 83:1004–1013

203. Hsu JC, Lin LC, Tzen JT, Chen JY (2011) Pardaxin-induced apoptosis enhances antitumor activity in HeLa cells. Peptides 32:1110–1116

204. Lelkes PI, Lazarovici P (1988) Pardaxin induces aggregation but not fusion of phosphatidylserine vesicles. FEBS Lett 230:131–136

205. Vad BS, Bertelsen K, Johansen CH, Pedersen JM, Skrydstrup T, Nielsen NC, Otzen DE (2010) Pardaxin permeabilizes vesicles more efficiently by pore formation than by disruption. Biophys J 98:576–585

206. Yang J, Parkanzky PD, Bodner ML, Duskin CA, Weliky DP (2002) Application of REDOR subtraction for filtered MAS observation of labeled backbone carbons of membrane-bound fusion peptides. J Magn Reson 159:101–110

207. Lee DK, Ramamoorthy A (1999) Determination of the solid-state conformations of polyalanine using magic-angle spinning NMR spectroscopy. J Phys Chem B 103:271–275

208. Batista CV, Scaloni A, Rigden DJ, Silva LR, Rodrigues Romero A, Dukor R, Sebben A, Talamo F, Bloch C (2001) A novel heterodimeric antimicrobial peptide from the treefrog Phyllomedusa distincta. FEBS Lett 494:85–89

209. Becucci L, Papini M, Mullen D, Scaloni A, Veglia G, Guidelli R (2011) Probing membrane permeabilization by the antimicrobial peptide distinctin in mercury-supported biomimetic membranes. Biochim Biophys Acta 1808:2745–2752

210. Cirioni O, Ghiselli R, Orlando F, Silvestri C, De Luca S, Salzano AM, Mocchegiani F, Saba

V, Scalise G, Scaloni A, Giacometti A (2008) Efficacy of the amphibian peptide distinctin in a neutropenic mouse model of staphylococcal sepsis. Crit Care Med 36:2629–2633

211. Raimondo D, Andreotti G, Saint N, Amodeo P, Renzone G, Sanseverino M, Zocchi I, Molle G, Motta A, Scaloni A (2005) A folding-dependent mechanism of antimicrobial peptide resistance to degradation unveiled by solution structure of distinctin. Proc Natl Acad Sci USA 102:6309–6314

212. Dalla Serra M, Cirioni O, Vitale RM, Renzone G, Coraiola M, Giacometti A, Potrich C, Baroni E, Guella G, Sanseverino M, De Luca S, Scalise G, Amodeo P, Scaloni A (2008) Structural features of distinctin affecting peptide biological and biochemical properties. Biochemistry 47:7888–7899

213. Shi L, Traaseth NJ, Verardi R, Cembran A, Gao J, Veglia G (2009) A refinement protocol to determine structure, topology, and depth of insertion of membrane proteins using hybrid solution and solid-state NMR restraints. J Biomol NMR 44:195–205

214. Bals R, Wang X, Zasloff M, Wilson JM (1998) The peptide antibiotic LL-37/hCAP-18 is expressed in epithelia of the human lung where it has broad antimicrobial activity at the airway surface. Proc Natl Acad Sci USA 95:9541–9546

215. Wong CC, Zhang L, Ren SX, Shen J, Chan RL, Cho CH (2011) Antibacterial peptides and gastrointestinal diseases. Curr Pharm Des 17:1583–1586

216. Wu WK, Wong CC, Li ZJ, Zhang L, Ren SX, Cho CH (2010) Cathelicidins in inflammation and tissue repair: potential therapeutic applications for gastrointestinal disorders. Acta Pharmacol Sin 31:1118–1122

217. Cederlund A, Gudmundsson GH, Agerberth B (2011) Antimicrobial peptides important in innate immunity. FEBS J 278:3942–3951

218. Tecle T, Tripathi S, Hartshorn KL (2010) Review: defensins and cathelicidins in lung immunity. Innate Immun 16:151–159

219. Yamasaki K, Gallo RL (2011) Rosacea as a disease of cathelicidins and skin innate immunity. J Investig Dermatol Symp Proc 15:12–15

220. Amatngalim GD, Nijnik A, Hiemstra PS, Hancock RE (2011) Cathelicidin peptide LL-37 modulates TREM-1 expression and inflammatory responses to microbial compounds. Inflammation 34:412–425

221. Brown KL, Poon GF, Birkenhead D, Pena OM, Falsafi R, Dahlgren C, Karlsson A, Bylund J, Hancock RE, Johnson P (2011) Host defense peptide LL-37 selectively reduces proinflammatory macrophage responses. J Immunol 186:5497–5505

222. Ramamoorthy A, Lee DK, Narasimhaswamy T, Nanga RP (2010) Cholesterol reduces pardaxin's dynamics-a barrel-stave mechanism of membrane disruption investigated by solid-state NMR. Biochim Biophys Acta 1798:223–227

223. Ramamoorthy A (2009) Beyond NMR spectra of antimicrobial peptides: dynamical images at atomic resolution and functional insights. Solid State Nucl Magn Reson 35:201–207

224. Hong J, Oren Z, Shai Y (1999) Structure and organization of hemolytic and nonhemolytic diastereomers of antimicrobial peptides in membranes. Biochemistry 38:16963–16973

225. Durr UH, Sudheendra US, Ramamoorthy A (2006) LL-37, the only human member of the cathelicidin family of antimicrobial peptides. Biochim Biophys Acta 1758:1408–1425

226. Henzler Wildman KA, Lee DK, Ramamoorthy A (2003) Mechanism of lipid bilayer disruption by the human antimicrobial peptide, LL-37. Biochemistry 42:6545–6558

227. Dubosclard V, Blondot ML, Eleouet JF, Bontems F, Sizun C (2011) 1H, 13C, and 15N resonance assignment of the central domain of hRSV transcription antitermination factor M2-1. Biomol NMR Assign 5:237–239

228. Aisenbrey C, Pendem N, Guichard G, Bechinger B (2012) Solid state NMR studies of oligourea foldamers: interaction of 15N-labelled amphiphilic helices with oriented lipid membranes. Org Biomol Chem 10:1440–1447

229. Pius J, Morrow MR, Booth V (2012) (2)H solid-state nuclear magnetic resonance investigation of whole Escherichia coli interacting with antimicrobial peptide MSI-78. Biochemistry 51:118–125

230. Bertelsen K, Vad B, Nielsen EH, Hansen SK, Skrydstrup T, Otzen DE, Vosegaard T, Nielsen NC (2012) Long-term-stable ether-lipid vs conventional ester-lipid bicelles in oriented solid-state NMR: altered structural information in studies of antimicrobial peptides. J Phys Chem B 115:1767–1774

Part IV

Engineering Approaches

Chapter 10

Prediction and Design of Outer Membrane Protein–Protein Interactions

Vikas Nanda, Daniel Hsieh, and Alexander Davis

Abstract

Protein–protein interactions (PPI) play central roles in biological processes, motivating us to understand the structural basis underlying affinity and specificity. In this chapter, we focus on biochemical and computational design strategies of assessing and detecting PPIs of β-barrel outer membrane proteins (OMPs). A few case studies are presented highlighting biochemical techniques used to dissect the energetics of oligomerization and determine amino acids forming the key interactions of the PPI sites. Current computational strategies for detecting/predicting PPIs are introduced, and examples of computational and rational engineering strategies applied to OMPs are presented.

Key words Protein–protein interactions, Outer membrane proteins, Computational design

Abbreviations

OMP	Outer membrane protein
OmpLA	Outer membrane phospholipase A
PPI	Protein–protein interaction
VDAC	Voltage-dependent anion channel

1 Introduction

Membrane proteins control a cell's interaction with its environment, and molecular interactions between proteins in the lipid bilayer serve to regulate these essential functions. A protein–protein interaction (PPI) is intended here to refer to an interface established by physical contacts between a pair of proteins that are specific and intentional in a biological context [1]. Such definition excludes general and transient binding sites such as those with ribosomes, chaperones, SRPs, and those implicated in the degradation pathway. Identifying protein partners, oligomerization state and the specific residues involved in the PPIs can help construct

Giovanna Ghirlanda and Alessandro Senes (eds.), *Membrane Proteins: Folding, Association, and Design*, Methods in Molecular Biology, vol. 1063, DOI 10.1007/978-1-62703-583-5_10, © Springer Science+Business Media, LLC 2013

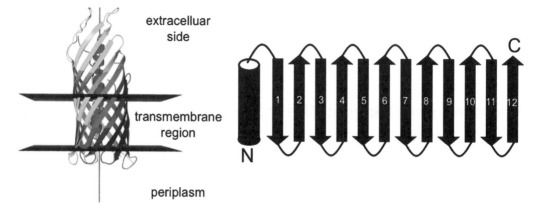

Fig. 1 Topology of an OMP. The bacterial autotransporter (PDB ID—1UYN) spans the lipid bilayer as a meandering beta sheet where strands adjacent in sequence form alternating antiparallel interactions

interactomes—system level maps of interactions between proteins [2]. Studying residues at PPIs can help us predict interaction networks and design experiments to verify the number and type of residues making the interactions. Furthermore, understanding the forces driving PPIs provides us insight in developing drugs that target these interaction sites.

Two major classes of membrane proteins are best described by their secondary structure: membrane-spanning α-helices and β-barrels. Transmembrane α-helical proteins are found almost exclusively in the inner membrane of Gram-negative bacteria, chloroplasts, and mitochondria [3, 4], whereas the transmembrane β-barrel OMPs are found primarily in the outer membrane with few known exceptions [5]. OMPs can have eight to twenty β-strands that weave back and forth through the outer membrane (Fig. 1). Some of these proteins can be found in active monomeric conformation, while others form functional oligomers. Most of the known oligomeric OMPs are trimeric and are involved in uptake of ions and nutrients via passive diffusion. Because oligomeric OMPs are essential to Gram-negative bacteria virulence, defense, and survival, it is crucial to understand the molecular mechanisms governing their higher-order structure.

Our current understanding of the intermolecular forces that guide the folding, insertion, and oligomerization of membrane proteins lags far behind that of water-soluble proteins. In particular, how this process is controlled in β-barrel OMPs is poorly understood. The same forces governing protein folding drive PPI formation: hydrophobic driving force, hydrogen bonding, electrostatics, and van der Waals interactions. PPIs in water-soluble proteins are often mediated by patches of hydrophobic residues on the protein surface, whose occlusion upon oligomerization stabilizes the complex [6]. In membrane proteins, the paradigm is inverted with polar interactions driving assembly [7, 8]. Even weakly polar

Fig. 2 Outer membrane phospholipase A. OmpLA is only catalytically active as a dimer (PDB ID 1QD6). Dimer formation is facilitated by the presence of calcium ions (*orange spheres*) and by the presence of substrate, mimicked here by the covalent sulfonylation of an active site residue. Few residues account for the PPI with the majority of intermolecular interactions between protein and substrate

interactions such as hydrogen bonds donated by the main chain Cα of glycine are believed to contribute to membrane protein stability [9–11]. Like α-helical membrane proteins, OMPs have a girdle of aromatic residues that interact with the headgroup region of the lipid bilayer [12].

Progress in understanding the insertion and folding of α-helical membrane proteins has benefitted greatly from model systems such as bacteriorhodopsin [13] and glycophorin A [14, 15] among others. Insights from these have led to the creation of de novo membrane proteins that stringently test models for membrane protein structure and folding [8]. Several model systems such as the outer membrane phospholipase A (OmpLA) now are providing similar insights for OMPs. OmpLA enzymatically modifies outer membrane lipids under stress conditions. Dimerization can be modulated in vitro by the addition of calcium [16] or by sulfonylation of active site residue Ser144 with a substrate analog [17] (Fig. 2). Using sedimentation equilibrium analytical ultracentrifugation (SE-AUC) to obtain an equilibrium consisting of monomers and dimers determined the monomer–dimer dissociation constants and resolved the apparent free energies for each condition [17]. Using calcium binding and sulfonylation to modulate dimerization thermodynamics, site-directed mutants of OmpLA have provided important insights into the molecular forces underlying OMP folding and PPI formation and how they often differ from α-helical membrane proteins. For example, the energetics of the self-association of bacteriorhodopsin and glycophorin A have been shown to depend on the magnitude of buried surface area [18, 19]. To test this, large amino acids were mutated to alanine to create cavities in the OmpLA PPI; no correlation was observed between the loss of surface area and the observed energetic changes upon mutation [20]. Similarly, inter-helical hydrogen bonding plays a significant role in folding of α-helical membrane proteins, where loss of a single hydrogen bond by mutation can change the oligomerization state or completely prevent association [7, 8, 21]. In OmpLA, alanine substitutions of residues forming hydrogen bonds across the PPI in high-resolution structures of the dimer had modest effects on complex stability [22]. Clearly the rules of α-helical membrane protein folding need to be reexamined in the context of OMP assembly.

Within OMPs themselves, there is significant variation in folding mechanisms and bilayer environmental preferences [23]. Other members of the OMP family are providing insight into the diversity of folding and oligomerization mechanisms. The mitochondrial voltage-dependent anion channel (VDAC) coordinates membrane permeability of ions and other biomolecules and may play a role in apoptosis [24–26]. The 19-stranded β-barrel is unusually plastic in that it has been shown to appear in dimeric, trimeric, and even hexameric arrangements [24, 27, 28]. Unlike OmpLA and many of the trimer forming porins, VDAC interfacial strands do not form a single, contiguous PPI. A structurally derived scoring function for identifying β-barrel strands involved in PPI formation was applied to VDAC [29]. Computationally designed mutations in these strands were shown to disrupt dimers or higher-order oligomers, helping dissect the multi-state oligomerization of this protein.

In the next section, several computational methods for predicting OMP PPI sites are presented that have been constructed using machine learning algorithms and knowledge-based potentials. Finally, several representative examples are discussed of engineered OMPs using both rational and computational methods.

2   Potentials for OMP Insertion and Oligomerization

Computational methods for predicting PPI sites in OMPs have many potential applications. Most methods do not require high-resolution structures, allowing PPI prediction from sequence-derived homology models. These methods can be used to identify oligomerization "hot spots," key residues involved in mediating protein interactions. They can be applied in the design of proteins with desired protein interaction properties. Methods for predicting OMP PPIs have been trained on available high-resolution structures in the Protein Data Bank [30]. There are over a hundred such structures available, but only a fraction of these exhibit realistic oligomer geometries within the crystal lattice or have conclusive biochemical characterizations of the oligomerization state and binding site residues [31]. As the number of high-quality experimental structures grows, so will the predictive power of the computational methods.

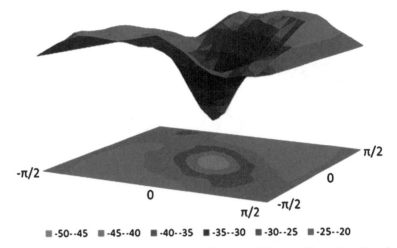

Fig. 3 Insertion landscape of OmpA. The Ezβ potential scores the orientation of OMPs based on the depth of lipid-exposed residues within the membrane bilayer. The *bottom* of the funnel (*red*) corresponds to the optimal orientation of OmpA where the barrel axis is nearly perpendicular to the membrane

The lipophilic surfaces of membrane proteins are optimized to interact with the continuously increasing environmental polarity from the hydrophobic inside of the lipid bilayer to the aqueous environments on either side. The distributions of each of the twenty amino acids as a function of membrane depth reflect their relative preferences for the aqueous, phospholipid headgroup and lipid acyl chain regions of the membrane. Using data sets of high-resolution structures, empirical functions describing these distributions were calculated for α-helical membrane proteins and OMPs [32–34]. The resulting $E_z\alpha$ and $E_z\beta$ potentials can be used to describe the energetics of protein insertion into the membrane. It is possible to visualize the energetic constraints on OMP orientation within the membrane as an "insertion funnel" where a vertical orientation in the membrane corresponds to optimal protein–lipid interactions as calculated by $E_z\beta$ (Fig. 3).

Residues involved in PPIs or specific interactions with other molecules such as lipopolysaccharides are often highly conserved and can be detected using molecular evolution-based methods [35]. Lipid-facing residues that function at such interfaces are not necessarily conserved for membrane insertion. Mapping the Ezβ energy of amino acids onto the surface of OMP structures shows a lower-than-expected frequency of depth compatible residues at PPI sites (Fig. 4). This bias can be quantitatively represented as an interaction moment whose magnitude and direction correlate with the interaction strength and PPI location. This approach can be applied to homology-derived structural models to predict PPI sites from protein sequence.

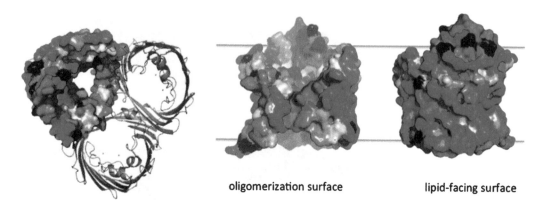

oligomerization surface   lipid-facing surface

Fig. 4 Identifying PPI promoting residues using Ezβ. The PPI of a trimeric porin (PDB ID—2POR) presents many residues with unfavorable lipid interaction propensities, whereas the lipid-facing side of the porin has predominantly favorable propensities. Residues are colored based on their normalized Ezβ score from *blue* (most favorable) to *red* (most unfavorable)

The *Transmembrane Strand Interaction Potential* (TMSIP) developed by Jackups and Liang [36] has also been developed to predict a number of features of OMPs from optimal strand pairing to stability and oligomerization state [37, 38]. This statistical potential was constructed using the β-barrel region of the OMP, classifying pairs of amino acids on adjacent strands based on their hydrogen-bonding environment. Highly overrepresented and underrepresented amino acid pairs gave physiochemical insights into both stabilizing and destabilizing features of membrane β-barrels. Tyrosine–glycine pairs were often found consistent with their known propensity to participate in an interaction called "aromatic rescue" [39], where the aromatic phenol group packs tightly against the glycine backbone, presumably shielding it from competing interactions that would destabilize the fold. Furthermore, a single-body term was developed, classifying the propensity of amino acids to localize to one of the five environments spanning the outer membrane—extracellular, extracellular headgroup, hydrophobic core, periplasmic headgroup, and periplasmic. Together, the single-body and pairwise terms effectively described the preferences for amino acids within the OMP.

TMSIP was subsequently extended to the prediction of OMP stability and oligomerization status [38]. It was found that simple, monomeric β-barrels had relatively high stabilities corresponding to low scores using the empirical potential. However, many barrels contain other ancillary structural elements, called out-clamps or in-plugs, depending on their location. Out-clamps are adjacent secondary structural elements that pack tightly against the barrel surface. In PagP, an amphipathic helix out-clamp lies along the outside edge of one end of the barrel (Fig. 5a). The 160-residue globular domain that inserts in the middle of the 22-stranded ferric

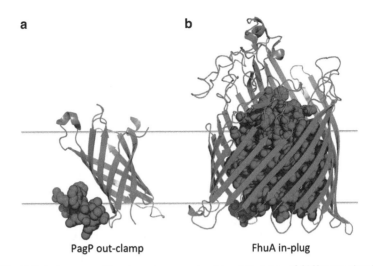

a   **b**

PagP out-clamp   FhuA in-plug

Fig. 5 Out-clamps and in-plugs. Large, extra-barrel domains of OMPs are classi-fied as (**a**) out-clamps and (**b**) in-plugs. These domains make extensive interac-tions with the transmembrane barrel, providing structural stability. OMPs with such domains are characterized by a high TMSIP score, indicating the barrels are unstable in the absence of the out-clamp or in-plug domain

hydroxide uptake receptor (FhuA) is an example of an in-clamp (Fig. 5b). In such cases, the stability as measured by TMSIP is con-sistently lower than that of simple, monomeric barrels. The lower stability of the β-barrel strands of OMPs is presumably compen-sated by favorable interactions with out-clamps and in-plugs. This is a common theme in soluble proteins, where protein interaction sites are intrinsically unstable and subsequently stabilized upon complex formation. This is of particular utility in cases where tran-sient, specific interactions are required for function. Binding affin-ity is attenuated by the free energy cost of stabilizing residues at the PPI, without compromising specificity of complex formation [40]. In many membrane proteins, molecular motions are required for conductance gating which presumably could be facilitated by transient interactions.

This strand-by-strand analysis has also been used to identify OMP oligomerization sites. In the case of out-clamps and in-plugs, the lower TMSIP scores are distributed across the entire barrel. However, oligomerization PPI sites often have significantly higher scores than the remainder of the barrel. Sites can therefore be iden-tified using both the average barrel stability as well as the variance in stability among strands. In some cases, OMPs that oligomerize contain in-plugs, out-clamps, or other specific biomolecule binding sites. In such cases, it can be challenging to identify PPI sites among the contributions of these other factors. The dimerization site of OmpLA, which is facilitated by calcium and/or substrate binding, is not discernable by TMSIP or $E_z\beta$ moment analysis [32, 38], which may account for why dimer stability is not dramatically affected by disrupting hydrogen-bonding networks or reducing occluded surface area at the PPI [17, 22].

Machine learning approaches have been utilized to identify which properties of lipid-facing OMP amino acids correlate most strongly with occurrence in PPIs. Numerous amino acid property scales have been developed over years of protein biophysical study, ranking amino acids according to polarity, size, α-helix propensity, and others. Over 500 such physiochemical indices have been compiled and made available online at AAINDEX [41]. The *Beta-barrel TransMembrane eXposure* (BTMX) server uses a combination of the best amino acid scales to identify buried versus lipid-facing amino acids and suggests which strands are involved in oligomeric interfaces [42]. Among the physical profiles that correlate most strongly with oligomerization are hydrophobicity and β-sheet propensity. It was found that oligomerization sites were more hydrophobic than other lipid-facing amino acids, which is perhaps surprising given the analysis was done on positions within the central 65 % of the lipid-facing positions. In contrast, lipid interaction potentials developed based on the statistical analysis of α-helical membrane protein structures indicate PPI sites tend to be less hydrophobic than lipid-facing sites [43]. It was also observed that strands participating in oligomerization tended to have a higher β-sheet forming propensity, which is in contrast to the TMSIP-derived model, where PPIs tend to have lower-than-average stability—suggesting that much of the predicted instability comes from pairwise interactions that are accounted for in TMSIP but not in BTMX. Discrepancies such as these in computational schemes for modeling OMPs highlight the importance of further methods development.

3   Engineered OMPs and Their Applications

OMPs are attractive targets for protein engineering because of their facility to form membrane-spanning pores. The β-barrel topology provides a central channel whose aperture is controlled to a large extent by the number of strands. The size of ligands conducted by these pores varies from ions to small molecules to other OMPs. There is great interest in repurposing these barrels as sensors where solutes are detected as they interrupt transmembrane electrical currents. Protein nanopores are finding applications in next-generation nucleic acid sequencing technologies and as reagents for controlled cell membrane permeabilization. Modulating chemical features of the pore by altering the protein sequence has the potential to provide exquisite ligand specificity. OMPs are often highly stable, such as α-hemolysin which remains folded in 7.2 M urea or at temperatures up to 93 °C [44, 45] and

OmpG which is stable even above the boiling point of water [46]. As a result, there have been many attempts to transplant OMPs into polymer membranes, nanoparticles, and solid supports for molecular sensing in nonbiological contexts.

Successful engineering of membrane protein structure and protein–protein interactions requires an adequate understanding of the intermolecular forces that govern folding and assembly in the membrane. Much of our knowledge of these forces has come from the biochemical study of engineered variants of membrane proteins or from bioinformatics methods such as those presented in the previous section. These principles of folding and assembly are stringently tested by the de novo design of membrane proteins. Hydrogen bonds and electrostatic interactions between polar amino acids in the lipid bilayer play a key role in mediating inter-helical interactions in the folding of monomeric α-helical transmembrane proteins and helical membrane protein oligomerization [7, 8]. A number of studies have exploited these interactions in altering the affinity or oligomerization state of protein–protein interactions among de novo model helical membrane protein [21, 47]. Analogous de novo model systems have yet to be developed for beta-barrel outer membrane proteins. As such, rational approaches to OMP design have focused on modifying the physical or chemical properties of existing natural OMPs. A few examples are selected representing the diverse rational engineering strategies used.

Due to the high degree of pore selectivity provided by OMPs, there has been great interest in incorporating these proteins into nonbiological environments that provide additional chemical stability or functionality. For example, the heptameric, 14-stranded barrel, α-hemolysin is an attractive target for nucleic acid sequencing as the size of its pore only allows passage of single-stranded DNA. α-Hemolysin chemically functionalized with a fragment of DNA was successfully incorporated into silicon wafers containing an array of 30 nm pores using the electrophoretic mobility of the nucleic acid to "thread" the protein into the solid-state nanopore when a voltage was applied across the wafer [48]. In this instance, minimal modifications to the protein were required to make it compatible with the silicon substrate. This is not always possible, as in the case of incorporation of the OMP FhuA into polymersomes, highly stable vesicles composed of block copolymers with hydrophobic and hydrophilic domains. The 22-stranded FhuA provides a large pore, over twice the diameter of α-hemolysin. The first 160 amino acids create a "cork" that blocks the channel pore. Removal of these amino acids creates a stable, unencumbered pore [49]. In natural proteins, the length of the nonpolar region of the lipid bilayer correlates with the thickness of the hydrophobic region of lipid-facing residues on membrane proteins. The difference in hydrophobic thickness of α-helical inner membrane proteins and OMPs is consistent with the lipid composition of these environments [32]. The increased thickness of the nonpolar region in polymersomes results in hydrophobic mismatch to OMPs, preventing their effective incorporation. In the case of an engineered FhuA nanochannel, it was necessary to increase the hydrophobic thickness by one nanometer through duplicating five amino acids of each of the 22 strands [50]. This "copy-and-paste" strategy resulted in a functional nanochannel in polymersomes, giving some appreciation of the impressive structural plasticity of OMPs.

The cellular location of OMPs makes them attractive targets for molecular engineering. These proteins are trafficked to and anchored in the outer membrane of Gram-negative bacteria like *E. coli*, where they present interstrand loops to the extracellular environment. The advent of protein display technology on the surface of microorganisms has the potential to create bacterial vaccines and cell-based reagents for biosensing, catalysis, and diagnostics [51]. Laboratory-directed evolution strategies have been successfully applied to alter substrate specificity of the *E. coli* endopeptidase OmpT [52]. Using fluorescent donor–quencher conjugated peptide substrates screened by flow cytometry, it was possible to identify mutants within a library of OmpT variants for activity against a novel target and lack of activity against the original OmpT target sequence. State-of-the-art OMP engineering consists mostly of rational or directed evolution approaches such as the examples presented. Computational design of membrane proteins is still a largely unexplored area, particularly in the case of OMP-like beta-barrels.

Computationally designed α-helical membrane proteins have been constructed using a combination of knowledge-based solvent interaction potentials and molecular mechanics tools. The redesign of the α-helical potassium channel KcsA into a water-soluble protein involved the simultaneous replacement of 140 lipid-facing residues [53]. In addition to typical intermolecular forces such as electrostatics and van der Waals interactions between amino acids, a key component of the scoring function was an "environmental energy" that described the hydrophobicity of exposed positions [54]. Introducing polar amino acids at these positions reduced the total hydrophobicity such that it would match a typical water-soluble protein of equivalent size. The resulting water-soluble design bound the same channel toxins that inactivate KcsA, indicating the three-dimensional structure was preserved despite such extensive mutagenesis. A solution NMR structure of the design subsequently validated that the full channel fold was maintained [55]. Another example of successful application of computational methods in the membrane was the design of α-helical peptides that bound to the transmembrane region of platelet integrins, disrupting protein–protein interactions that modulated thrombosis [56]. The lack at the time of high-resolution structures for transmembrane integrin domains made this particularly challenging, as the number of degrees of freedom in main chain conformation and inter-protein interactions is enormous. To effectively sample this large conformation space, the designs were constructed on a library of helix–helix interactions extracted from existing protein structures with the assumption that this represented the energetically relevant subset of available conformations. Integrin target sequences were threaded onto one helix of the pair, while remaining helix sequence was optimized for molecular packing forces and a knowledge-based lipid interaction potential. The resulting peptide sequences showed high affinity and specificity for their target integrin. These examples have served to benchmark existing design methods that are now being extended to α-helical transmembrane proteins that conjugate nonnatural cofactors or protein mimetics using beta-amino acids that target protein–protein interactions [57, 58].

Examples of computational tools to the design of OMPs are only beginning to emerge. In a study to explore the use of outer membrane protein G (OmpG) as an alternative for α-hemolysin in analyte detection, variants of the OmpG barrel were designed which reduced the flexibility of external loops, lowering spontaneous gating that would contribute significant background noise [59]. This was accomplished by deleting an amino acid that caused one of the strands to "bulge," disrupting the network of backbone hydrogen bonds that stabilize the barrel structure. Further stabilization was achieved by introducing a disulfide bond into the extracellular loops. The structural consequences of these modifications were evaluated using all-atom molecular dynamics, which models molecular motions on the picosecond to nanosecond time scale. Simulations showed that the modifications reduced loop mobility and improved interstrand hydrogen bond occupancy. The OmpG porin with stabilizing modifications showed a 95 % reduction in spontaneous gating. In this case, molecular dynamics used to evaluate the effect of designed variants arrived at by inspection of experimentally high-resolution structures. Using molecular dynamics to suggest variants is more challenging due to the computational expense associated with all-atom simulations, particularly in the context of a modeled lipid bilayer.

The oligomerization state of Oomph was modulated using a computational approach to direct the choice of mutations [60]. OmpF naturally exists as a trimer although a dimeric form has also been observed under certain conditions. Using the TMSIP potential discussed in the previous section [38], a set of lipid-facing amino acids were identified as most responsible for the instability of strands located at the trimer interface. Several of these amino acids were glycines involved in tight interchain packing at the protein–protein interface. Mutations to small amino acids such as alanine or serine had modest affects on oligomer stability, whereas leucine/isoleucine substitutions clearly disrupted oligomerization. The TMSIP potential was successful in identifying oligomerization hot spots. Interestingly, many of the mutations tested disrupted oligomer formation without impacting monomer stability, suggesting a two-state model of folding followed by insertion may hold for OMP oligomers, much as it does of α-helical membrane proteins [13]. This same potential has been extended to identify key residues involved in the oligomerization of the mitochondrial VDAC1 [29] and to enhance stability of the structurally related human Tom40 protein [61].

Rational engineering methods combined with emerging computational tools have the potential to facilitate OMP engineering for medical and biotechnological applications. They also improve our basic understanding of OMP folding, oligomerization, and stability.

References

1. De Las Rivas J, Fontanillo C (2010) Protein-protein interactions essentials: key concepts to building and analyzing interactome networks. PLoS Comput Biol 6(6):e1000807

2. Cusick ME et al (2005) Interactome: gateway into systems biology. Hum Mol Genet 14(Spec No. 2):R171–R181

3. Schleiff E et al (2003) Characterization of the translocon of the outer envelope of chloroplasts. J Cell Biol 160(4):541–551

4. Schein SJ, Colombini M, Finkelstein A (1976) Reconstitution in planar lipid bilayers of a voltage-dependent anion-selective channel obtained from paramecium mitochondria. J Membr Biol 30(2):99–120

5. Dong C et al (2006) Wza the translocon for E. coli capsular polysaccharides defines a new class of membrane protein. Nature 444(7116):226–229

6. Jones S, Thornton JM (1997) Prediction of protein-protein interaction sites using patch analysis. J Mol Biol 272(1):133–143

7. Zhou FX et al (2000) Interhelical hydrogen bonding drives strong interactions in membrane proteins. Nat Struct Biol 7(2):154–160

8. Choma C et al (2000) Asparagine-mediated self-association of a model transmembrane helix. Nat Struct Biol 7(2):161–166

9. Senes A, Engel DE, DeGrado WF (2004) Folding of helical membrane proteins: the role of polar, GxxxG-like and proline motifs. Curr Opin Struct Biol 14(4):465–479

10. Senes A, Ubarretxena-Belandia I, Engelman DM (2001) The Calpha —H...O hydrogen bond: a determinant of stability and specificity in transmembrane helix interactions. Proc Natl Acad Sci USA 98(16):9056–9061

11. Senes A, Gerstein M, Engelman DM (2000) Statistical analysis of amino acid patterns in transmembrane helices: the GxxxG motif occurs frequently and in association with beta-branched residues at neighboring positions. J Mol Biol 296(3):921–936

12. Schulz GE (2002) The structure of bacterial outer membrane proteins. Biochim Biophys Acta 1565(2):308–317

13. Popot JL, Engelman DM (1990) Membrane protein folding and oligomerization: the two-stage model. Biochemistry 29(17):4031–4037

14. Lemmon MA et al (1992) Sequence specificity in the dimerization of transmembrane alpha-helices. Biochemistry 31(51):12719–12725

15. Lemmon MA et al (1992) Glycophorin A dimerization is driven by specific interactions between transmembrane alpha-helices. J Biol Chem 267(11):7683–7689

16. Dekker N et al (1997) Dimerization regulates the enzymatic activity of Escherichia coli outer membrane phospholipase A. J Biol Chem 272(6):3179–3184

17. Stanley AM et al (2006) Energetics of outer membrane phospholipase A (OMPLA) dimerization. J Mol Biol 358(1):120–131

18. Doura AK et al (2004) Sequence context modulates the stability of a GxxxG-mediated transmembrane helix-helix dimer. J Mol Biol 341(4):991–998

19. Faham S et al (2004) Side-chain contributions to membrane protein structure and stability. J Mol Biol 335(1):297–305

20. Ebie Tan A, Fleming KG (2008) Outer membrane phospholipase a dimer stability does not correlate to occluded surface area. Biochemistry 47(46):12095–12103

21. Cristian L et al (2005) Synergistic interactions between aqueous and membrane domains of a designed protein determine its fold and stability. J Mol Biol 348(5):1225–1233

22. Stanley AM, Fleming KG (2007) The role of a hydrogen bonding network in the

transmembrane beta-barrel OMPLA. J Mol Biol 370(5): 912–924

23. Burgess NK et al (2008) Beta-barrel proteins that reside in the Escherichia coli outer membrane in vivo demonstrate varied folding behavior in vitro. J Biol Chem 283(39):26748–26758

24. Hoogenboom BW et al (2007) The supramolecular assemblies of voltage-dependent anion channels in the native membrane. J Mol Biol 370(2):246–255

25. Granville DJ, Gottlieb RA (2003) The mitochondrial voltage-dependent anion channel (VDAC) as a therapeutic target for initiating cell death. Curr Med Chem 10(16): 1527–1533

26. Shoshan-Barmatz V, Gincel D (2003) The voltage-dependent anion channel: characterization, modulation, and role in mitochondrial function in cell life and death. Cell Biochem Biophys 39(3):279–292

27. Zalk R et al (2005) Oligomeric states of the voltage-dependent anion channel and cytochrome c release from mitochondria. Biochem J 386(Pt 1):73–83

28. Hiller S et al (2008) Solution structure of the integral human membrane protein VDAC-1 in detergent micelles. Science 321(5893): 1206–1210

29. Geula S et al (2012) Structure-based analysis of VDAC1 protein: defining oligomer contact sites. J Biol Chem 287(3):2179–2190

30. Protein Data Bank. http://www.rcsb.org/pdb.

31. Fairman JW, Noinaj N, Buchanan SK (2011) The structural biology of beta-barrel membrane proteins: a summary of recent reports. Curr Opin Struct Biol 21(4):523–531

32. Hsieh D, Davis A, Nanda V (2012) A knowledge-based potential highlights unique features of membrane alpha-helical and beta-barrel protein insertion and folding. Protein Sci 21(1):50–62

33. Senes A et al (2007) E(z), a depth-dependent potential for assessing the energies of insertion of amino acid side-chains into membranes: derivation and applications to determining the orientation of transmembrane and interfacial helices. J Mol Biol 366(2):436–448

34. Schramm CA et al (2012) Knowledge-based potential for positioning membrane-associated structures and assessing residue-specific energetic contributions. Structure 20(5):924–935

35. Adamian L, Naveed H, Liang J (2011) Lipid-binding surfaces of membrane proteins: evidence from evolutionary and structural analysis. Biochim Biophys Acta 1808(4): 1092–1102

36. Jackups R Jr, Liang J (2005) Interstrand pairing patterns in beta-barrel membrane proteins: the positive-outside rule, aromatic rescue, and strand registration prediction. J Mol Biol 354(4):979–993

37. Naveed H et al (2012) Predicting three-dimensional structures of transmembrane domains of beta-barrel membrane proteins. J Am Chem Soc 134(3):1775–1781

38. Naveed H, Jackups R Jr, Liang J (2009) Predicting weakly stable regions, oligomerization state, and protein-protein interfaces in transmembrane domains of outer membrane proteins. Proc Natl Acad Sci USA 106(31): 12735–12740

39. Merkel JS, Regan L (1998) Aromatic rescue of glycine in beta sheets. Fold Des 3(6): 449–455

40. Dyson HJ, Wright PE (2005) Intrinsically unstructured proteins and their functions. Nat Rev Mol Cell Biol 6(3):197–208

41. AAINDEX. http://www.genome.jp/aaindex.

42. Hayat S et al (2011) Prediction of the exposure status of transmembrane beta barrel residues from protein sequence. J Bioinforma Comput Biol 9(1):43–65

43. Adamian L et al (2005) Empirical lipid propensities of amino acid residues in multispan alpha helical membrane proteins. Proteins 59(3):496–509

44. Jung Y, Bayley H, Movileanu L (2006) Temperature-responsive protein pores. J Am Chem Soc 128(47):15332–15340

45. Gu LQ et al (1999) Stochastic sensing of organic analytes by a pore-forming protein containing a molecular adapter. Nature 398(6729):686–690

46. Korkmaz-Ozkan F et al (2010) Correlation between the OmpG secondary structure and its pH-dependent alterations monitored by FTIR. J Mol Biol 401(1):56–67

47. Tatko CD et al (2006) Polar networks control oligomeric assembly in membranes. J Am Chem Soc 128(13):4170–4171

48. Hall AR et al (2010) Hybrid pore formation by directed insertion of alpha-haemolysin into solid-state nanopores. Nat Nanotechnol 5(12):874–877

49. Mohammad MM, Howard KR, Movileanu L (2011) Redesign of a plugged beta-barrel membrane protein. J Biol Chem 286(10): 8000–8013

50. Muhammad N et al (2011) Engineering of the E. coli outer membrane protein FhuA to overcome the hydrophobic mismatch in thick polymeric membranes. J Nanobiotechnology 9:8

51. Georgiou G et al (1997) Display of heterologous proteins on the surface of microorganisms: from the screening of combinatorial libraries to live recombinant vaccines. Nat Biotechnol 15(1):29–34

52. Varadarajan N et al (2008) Highly active and selective endopeptidases with programmed substrate specificities. Nat Chem Biol 4(5):290–294

53. Slovic AM et al (2004) Computational design of water-soluble analogues of the potassium channel KcsA. Proc Natl Acad Sci USA 101(7):1828–1833

54. Kono H, Saven JG (2001) Statistical theory for protein combinatorial libraries. Packing interactions, backbone flexibility, and the sequence variability of a main-chain structure. J Mol Biol 306(3):607–628

55. Ma D et al (2008) NMR studies of a channel protein without membranes: structure and dynamics of water-solubilized KcsA. Proc Natl Acad Sci USA 105(43): 16537–16542

56. Yin H et al (2007) Computational design of peptides that target transmembrane helices. Science 315(5820):1817–1822

57. Shandler SJ et al (2011) Computational design of a beta-peptide that targets transmembrane helices. J Am Chem Soc 133(32): 12378–12381

58. Korendovych IV et al (2010) De novo design and molecular assembly of a transmembrane diporphyrin-binding protein complex. J Am Chem Soc 132(44):15516–15518

59. Chen M et al (2008) Outer membrane protein G: engineering a quiet pore for biosensing. Proc Natl Acad Sci U S A 105(17):6272–6277

60. Naveed H et al (2012) Engineered oligomerization state of OmpF protein through computational design decouples oligomer dissociation from unfolding. J Mol Biol 419(1–2):89–101

61. Gessmann D et al (2011) Improving the resistance of a eukaryotic beta-barrel protein to thermal and chemical perturbations. J Mol Biol 413(1):150–161

Chapter 11

Design of Transmembrane Peptides: Coping with Sticky Situations

Arianna Rath and Charles M. Deber

Abstract

Membrane proteins have central roles in cellular processes ranging from nutrient uptake to cell–cell communication, and are key drug targets. However, research on α-helical integral membrane proteins is in its relative infancy vs. water-soluble proteins, largely because of their water insolubility when extracted from their native membrane environment. Peptides with sequences that correspond to the membrane-spanning segments of α-helical integral membrane proteins, termed transmembrane (TM) peptides, provide valuable tools for the characterization of these molecules. Here we describe in detail protocols for the design of TM peptides from the sequences of natural α-helical integral membrane proteins and outline strategies for their synthesis and for improving their solubility properties.

Key words Alpha-helical integral membrane protein, Transmembrane sequence, Peptide design, Peptide synthesis, Solubility

1 Introduction

Alpha-helical integral membrane proteins are vital to cellular processes ranging from nutrient uptake to cell–cell communication, and constitute more than half of drug targets [1, 2]. Efforts to understand and control their structures are understandably of great interest in both basic and applied research. Yet, despite advances in their isolation and structural characterization, α-helical membrane proteins currently represent only 1–2 % of the >80,000 structures deposited in the Protein Data Bank, largely because of the technical challenges associated with the over-expression, purification, and structural characterization of these highly hydrophobic molecules.

Insight into α-helical membrane protein folding has accordingly not relied exclusively on high-resolution structure determination. Consideration of the individual transmembrane (TM) α-helices as independently folded units has facilitated a "divide and conquer" approach to the study of membrane protein assembly, where interactions between peptides that recapitulate the

Giovanna Ghirlanda and Alessandro Senes (eds.), *Membrane Proteins: Folding, Association, and Design*,
Methods in Molecular Biology, vol. 1063, DOI 10.1007/978-1-62703-583-5_11, © Springer Science+Business Media, LLC 2013

membrane-embedded portions of α-helical membrane proteins can be studied in vitro and in vivo (reviewed in ref. 3). These peptides correspond to, or are derived from, one or more natural TM sequences, defined as membrane-spanning contiguous strands (~20–30 amino acids, *see* ref. 4) of ~80 % hydrophobic residues (e.g., Leu, Ile, Val), often flanked at each end by aromatic and/or positively charged residues at the bilayer periphery [4]. Here we describe the design and optimization of peptide sequences from the membrane-spanning sequences of single- or multi-spanning α-helical integral membrane proteins, termed "TM peptides," and outline procedures to aid their reconstitution into micellar media.

2   Materials

Aqueous solutions should be made with ultrapure water (prepared by purifying deionized water to attain a sensitivity of 18 M Ω cm at 25 °C). Use reagent grade chemicals and room temperature storage unless otherwise indicated. A fume hood and appropriate personal protective equipment (lab coat, gloves, and goggles) are required for handling of the corrosive and volatile materials used in the peptide reconstitution procedure. Adhere to appropriate regulations for disposal of waste materials.

2.1 Identification and Optimization of TM Peptide Sequences

1. Personal computer and Internet access are required. TM prediction web servers are free for academic users; fees may apply for nonacademic users in certain instances.

2.2 Solubilization of TM Peptides

1. 1,1,1,3,3,3-Hexafluoro-2-propanol (HFIP), reagent plus, ≥99 %.
2. Trifluoroacetic acid (TFA), reagent grade, 99 %.
3. 2,2,2-Trifluoroethanol (TFE), reagent plus, ≥99 %.
4. Dry ice pellets and laboratory grade ethanol.
5. Lyophilizer maintained at or below −70 °C.
6. Glass screw cap vials and screw caps with PTFE liner seals.
7. Compressed N_2 gas stream accessible within a fume hood.
8. Solution of appropriate high-purity detergent in ultrapure water.

3   Methods

3.1 Transmembrane Sequence Identification

If structural information of sufficient resolution is available, the start and end sites of membrane-embedded helical structures may be used to determine the peptide sequence of the TM segment of interest. Biochemical investigations of accessibility of engineered or native Cys residues to membrane-permeable vs. membrane-impermeable modifying reagents can also aid selection of TM vs. non-TM residues but may label the pore-lining residues of permeases as water-exposed. For many membrane proteins, however, structural and biochemical investigations of topology are not available. In these cases, or to complement existing data, selection of residues that should be incorporated in a TM peptide sequence may be guided by computational methods, as follows:

1. Obtain the amino acid sequence of the protein of interest, ideally in FASTA format. If this format is not available, use a text file of single-letter abbreviations for amino acids.

2. Submit the sequence to at least 10–15 TM prediction software Web servers that provide per-residue prediction (*see* Table 1 for a list of servers and URLs). This can be done in a cut-and-paste manner or by uploading a FASTA file in certain instances. Select from algorithms based on different principles and trained on different learning sets (*see* **Notes 1** and **2**).

3. Examine the predicted TM sequences for consensus. Those residues identified in at least one-half of predictions as TM should be included in the TM peptide sequence (*see* Fig. 1 and **Note 2**).

3.2 Lys-Tag Incorporation

The high hydrophobicity that typifies membrane-embedded sequences often results in peptides that are poorly soluble in water, leading to challenges in their synthesis, purification, and/or resolubilization in membrane-mimetic media. Lys-tagging, defined as the addition of Lys residues at the N- and/or C-terminus of the TM peptide sequence, provides a means of enhancing the water solubility and minimizing the aggregation propensity of TM peptides that is generally compatible with retention of native self-assembly and structure propensities [5–10]. The utility of Lys tags lies in their ionization at biomimetic pH values, which both increases solubility and destabilizes peptide aggregation via unfavorable electrostatic interactions. Incorporation of Lys tags into TM peptide sequences may be performed as follows:

1. Calculate the total Liu–Deber hydrophobicity score [31] of the TM peptide based on the frequency of occurrence of each residue in its sequence (*see* Table 2 for values).

2. Add the minimum number of Lys residues needed to reduce the total Liu–Deber hydrophobicity score of the peptide sequence to a value <10 (*see* **Note 3**). Peptides with scores below this threshold are typically sufficiently soluble in water for routine purification and characterization in aqueous, detergent-free, buffers (*see* **Note 4**).

3. Determine the expected net charge of the peptide at neutral pH. Adjust number of Lys tags and/or peptide sequence such that the TM peptide carries a nonzero net charge (*see* **Note 4**). Consider Asp and Glu to carry a net charge of −1; Lys and Arg +1;

Table 1
TM sequence prediction methods[a]

Name	Reference	URL/description[b]
DAS-TMFilter	[32]	http://mendel.imp.ac.at/sat/DAS/DAS.html Dense alignment surface method compares query sequence to collection of nonhomologous membrane proteins for prediction
HMMTOP 2.0	[33]	http://www.enzim.hu/hmmtop/ Unsupervised machine learning based on 5-state (inside loop, inside helix tail, membrane helix, outside helix tail, and outside loop) hidden Markov model
MemBrain	[34]	http://www.csbio.sjtu.edu.cn/bioinf/MemBrain/ Machine learning integrating various bioinformatics approaches including multiple sequence alignment, the optimized evidence-theoretic K-nearest neighbor prediction algorithm, fusion of multiple prediction window sizes, and classification by dynamic threshold
MemPype (ENSEMBLE)	[35]	http://mu2py.biocomp.unibo.it/mempype Combination of cascade-neural network, two hidden Markov models, trained on 59 high-resolution helical membrane protein structures
MEMSAT3	[36]	http://bioinf.cs.ucl.ac.uk/psipred/ Model recognition approach based on amino acid preferences taken from various structural parts of membrane proteins
MEMSAT-SVM	[37]	http://bioinf.cs.ucl.ac.uk/psipred/ Support vector machine-based prediction based on MEMSAT3 sequence profiles
MPEX	[38]	http://blanco.biomol.uci.edu/mpex/ "Sliding window" based on experimentally determined water–octanol and water–liposome partitioning coefficients of amino acid residues
OCTOPUS[c,d]	[39]	http://topcons.cbr.su.se/index.php Combines hidden Markov models and artificial neural networks. Incorporates modeling of reentrant loops/membrane-dipping regions
OrienTM	[40]	http://athina.biol.uoa.gr/orienTM/ Simple algorithm that identifies TMs based on position-specific parameters for residues which belong to non-TM segments derived from all non-TM sequences in the SwissProt database
PHDhtm	[41]	http://npsa-pbil.ibcp.fr/cgi-bin/npsa_automat.pl?page=/NPSA/npsa_htm.html Neural network system trained on multiple sequence alignments
Philius	[42]	http://noble.gs.washington.edu/proj/philius/ Dynamic Bayesian network method that combines a signal peptide submodel with a TM submodel
Phobius	[43]	http://phobius.cgb.ki.se/ Delineates signal peptide and TM regions using hidden Markov model (HMM) and homology information. Trained on a newly assembled and curated dataset

(continued)

Table 1
(continued)

PRO[c]	[44]	http://topcons.cbr.su.se/index.php Hidden Markov model based on evolutionary information profiles derived from multiple sequence alignments
PRODIV[c]	[44]	http://topcons.cbr.su.se/index.php Hidden Markov model based on evolutionary information profiles derived from multiple sequence alignments
SCAMPI-seq[c,e]	[45]	http://scampi.cbr.su.se/ Single-sequence version of hidden Markov model based on an experimental scale of position-specific amino acid contributions to the free energy of membrane insertion
SCAMPI-msa[c]	[45]	http://topcons.cbr.su.se/index.php Multiple sequence version of SCAMPI
SOSUI	[46]	http://bp.nuap.nagoya-u.ac.jp/sosui/sosui_submit.html "Sliding window" based on four amino acid physicochemical parameters, the average hydrophobicity, and the average amphiphilicity index in the end region of helices
SPLIT4	[47]	http://split.pmfst.hr/split/4/ Preference function method based on 15 amino acid physicochemical property scales, incorporating positive charge distribution
SPOCTOPUS	[48]	http://octopus.cbr.su.se/ Pre-processor to OCTOPUS that uses a neural network and hidden Markov model to detect and locate, respectively, a signal peptide in the 70 most N-terminal residues of a sequence
SVMtop	[49]	http://bio-cluster.iis.sinica.edu.tw/~bioapp/SVMtop/index_adv.php Hierarchical support vector machine method that distinguishes TM and non-TM residues by incorporating several biological features of a TM helix in a lipid bilayer environment, then evaluates topology with a scoring function based on membrane protein folding
TM Finder	[50]	http://tmfinder.research.sickkids.ca/cgi-bin/TMFinderForm.cgi "Sliding window" based on experimentally determined amino acid residue helix propensity in nonpolar solvent and the Liu–Deber hydrophobicity scale (*see* Table 2)
TMHMM	[51]	http://www.cbs.dtu.dk/services/TMHMM/ Supervised machine learning using 7-state (globular, cytoplasmic loop, cytoplasmic cap, helix core, non-cytoplasmic short loop, non-cytoplasmic long loop, non-cytoplasmic cap) hidden Markov model
TMPro	[52]	http://linzer.blm.cs.cmu.edu/tmpro Neural network trained to predict TM or non-TM using multiple amino acid properties (charge, polarity, aromaticity, size, and electronic properties) extracted from sequence information by applying the framework used for latent semantic analysis
TopPred	[53]	http://bioweb.pasteur.fr/seqanal/interfaces/toppred.html "Sliding window" based on hydrophobicity and positive charge distribution

[a] A non-comprehensive list of TM helix prediction Web servers is provided in alphabetical order. Independent assessments of TM helix prediction accuracy may be found in refs. 14, 28

[b] The general approaches to prediction are reviewed in ref. 13, 29

[c] These prediction methods are run concurrently by the TOPCONS web server [30]

[d] OCTOPUS is available as part of TOPCONS and at the SPOCTOPUS web server

[e] SCAMPI-seq is available as part of TOPCONS and at the indicated URL

Prediction software	58	59	60	61	62	63	64	65	66	67	68	69	70	71	72	73	74	75	76	77	78	79	80	81	82	83	84	85	86	87	88	89	90	91	92	93	94	95	96
Full-length PLP sequence	E	Y	L	I	N	V	I	H	A	F	Q	Y	V	I	Y	G	T	A	S	F	F	F	L	Y	G	A	L	L	A	E	G	F	Y	T	T	G	A	V	
DAS-TMFilter										F	Q	Y	V	I	Y	G	T	A	S	F	F	F	L	Y	G	A	L	L	A	E	G								
HMMTOP 2.0												Y	V	I	Y	G	T	A	S	F	F	F	L	Y	G	A	L	L	A	E	G	F	Y	T	T				
MemBrain					N	V	I	H	A	F	Q	Y	V	I	Y	G	T	A	S	F	F	F	L	Y	G	A	L	L	A	E	G	F	Y	T					
MemPype (ENSEMBLE)			L	I	N	V	I	H	A	F	Q	Y	V	I	Y	G	T	A	S	F	F	F	L	Y	G	A	L	L	A	E	G	F	Y						
MEMSAT3										F	Q	Y	V	I	Y	G	T	A	S	F	F	F	L	Y	G	A	L	L	A	E	G	F	Y						
MEMSAT-SVM					N	V	I	H	A	F	Q	Y	V	I	Y	G	T	A	S	F	F	F	L	Y	G	A	L	L	A	E	G	F	Y	T					
MPEX						V	I	H	A	F	Q	Y	V	I	Y	G	T	A	S	F	F	F	L	Y	G	A	L	L	A	E	G	F	Y	T					
OCTOPUS																	T	A	S	F	F	F	L	Y	G	A	L	L	A	E	G	F	Y	T	T	G			
OrienTM												Y	V	I	Y	G	T	A	S	F	F	F	L	Y	G	A	L	L	A	E	G	F	Y	T					
PHDhtm						V	I	H	A	F	Q	Y	V	I	Y	G	T	A	S	F	F	F	L	Y	G	A	L	L	A	E	G	F	Y	T					
Philius											Q	Y	V	I	Y	G	T	A	S	F	F	F	L	Y	G	A	L	L	A	E	G	F	Y						
Phobius		Y	L	I	N	V	I	H	A	F	Q	Y	V	I	Y	G	T	A	S	F	F	F	L	Y	G	A	L	L	A										
PRO										F	Q	Y	V	I	Y	G	T	A	S	F	F	F	L	Y	G	A	L	L	A	E	G	F	Y	T	T				
PRODIV																G	T	A	S	F	F	F	L	Y	G	A	L	L	A	E	G	F	Y	T					
SCAMPI-seq															Y	G	T	A	S	F	F	F	L	Y	G	A	L	L	A	E	G	F	Y	T					
SCAMPI-msa																	T	A	S	F	F	F	L	Y	G	A	L	L	A	E	G	F	Y	T	T	G			
SOSUI									A	F	Q	Y	V	I	Y	G	T	A	S	F	F	F	L	Y	G	A	L	L	A	E	G	F	Y	T					
SPLIT4									A	F	Q	Y	V	I	Y	G	T	A	S	F	F	F	L	Y	G	A	L	L	A	E	G	F							
SPOCTOPUS									A	F	Q	Y	V	I	Y	G	T	A	S	F	F	F	L	Y	G	A	L	L	A	E	G	F	Y	T					
SVMtop			L	I	N	V	I	H	A	F	Q	Y	V	I	Y	G	T	A	S	F	F	F	L	Y	G	A	L	L	A	E	G	F	Y	T					
TM Finder		Y	L	I	N	V	I	H	A	F	Q	Y	V	I	Y	G	T	A	S	F	F	F	L	Y	G	A	L	L	A	E	G	F	Y	T	T	G	A		
TMHMM		Y	L	I	N	V	I	H	A	F	Q	Y	V	I	Y	G	T	A	S	F	F	F	L	Y	G	A	L	L	A										
TMPro		Y	L	I	N	V	I	H	A	F	Q	Y	V	I	Y	G	T	A	S	F	F	F	L	Y	G	A	L	L	A	E	G								
TopPred												Y	V	I	Y	G	T	A	S	F	F	F	L	Y	G	A	L	L	A	E	G								
Prediction frequency (n)	0	4	6	6	8	11	11	11	12	15	16	19	20	21	22	24	24	24	24	23	23	23	23	18	16	14	13	10	5	3	1	0							
Consensus prediction									A	F	Q	Y	V	I	Y	G	T	A	S	F	F	F	L	Y	G	A	L	L	A	E	G	F	Y						

Legend (n shading):
- 0 <= n < 4
- 4 <= n < 8
- 8 <= n < 12
- 12 <= n < 16
- 16 <= n < 20
- 20 <= n <= 24

Fig. 1 Selecting a TM peptide sequence by consensus prediction. TM segment 2 of an oligomeric, four-helix integral membrane protein involved in maintenance of the lamellar structure of central nervous system myelin [27], bovine myelin proteolipid protein (PLP), is used as an example. The complete sequence of PLP (UniProt ID P04116) was submitted to the 24 Web servers listed in Table 1 and TM predictions made using default settings. The resulting predictions of the second TM sequence (PLP TM2) are shown ordered from top to bottom according to the Web server list in Table 1. n indicates the number of times a given residue was predicted as TM and ranges from 0 (not identified as a TM residue among the 24 predictions) to 24 (identified as a TM residue in each prediction). The PLP sequence incorporated into the TM2 peptide is shown at the bottom. A TM2 peptide sequence that did not incorporate Phe 90 was incapable of forming oligomers in vitro; however, addition of both Phe 90 and Tyr 91 to the sequence had no effect on peptide self-assembly [15]. PLP sequence numbering begins after the initiator Met, following ref. 15

Table 2
Liu–Deber amino acid residue hydrophobicity values

Residue	Hydrophobicity[a]
A	0.17
C	2.49
D	−2.49
E	−1.49
F	5.00
G	−3.31
H	−4.63
I	4.41
K	−5.00
L	4.76
M	3.23
N	−3.79
P	−4.92
Q	−2.75
R	−2.77
S	−2.84
T	−1.08
V	3.02
W	4.88
Y	2.00

[a]Relative hydrophobicity of amino acid residues determined by RP-HPLC retention times of peptides with the general sequence KK-AAAXAAAAAXAAWAAXAAA-KKKK, where "**X**" is each of the indicated residues. Data adapted from ref. 31

His zero; an amino N-terminus +1, an acylated N-terminus zero, an amidated peptide C-terminus zero, and an acidic peptide C-terminus −1 (*see* **Note 5**).

4. Decide on the N- and/or C-terminal distribution of Lys residues. Symmetric placement at the N- and C-terminal ends of the peptide sequence is ideal for water solubility. For odd numbers of Lys residues, place the odd Lys at the peptide terminus predicted to be cytoplasmic (*see* **Note 6**). Certain peptide sequences and/or applications may require asymmetric, all-N, or all-C localization of Lys residues and/or utilization of alternate polar residue tags (*see* **Note 7**). We additionally caution that Lys-tagging may alter peptide oligomerization in certain instances (*see* **Note 8**).

3.3 Sequence Optimization

At this point, the desired TM peptide sequence will most likely consist of a central segment that corresponds to the natural protein sequence, flanked on both sides by Lys residues. Synthesis of the natural TM helix sequence is ideal, particularly if the impact of residue substitutions in the full-length protein is unknown. Certain sequence adjustments may nevertheless be desired for peptide synthesis and work-up. Below we describe some TM peptide sequence alterations that may be useful. The reader is referred elsewhere for a description of solid-phase peptide synthesis (SPPS) and peptide cleavage/deprotection strategies [11].

1. Inspect the TM peptide sequence for Cys and Met. Substitute with an appropriate residue(s) if reducing agents will interfere with downstream peptide applications (*see* **Note 9**).

2. Avoid use of Asn or Gln as the N-terminal residue, Pro and/or Gly in the two positions preceding the C-terminus, and eliminate Asp–Asn, Asp–Ala, Asp–Gly, Asp–Pro, or Asp–Ser pairs if possible ([11], *see* **Note 10**).

3.4 Solubilization of TM Peptides

Under ideal circumstances, crude and purified Lys-tagged TM peptides will readily dissolve in pure water. However, even the best design efforts occasionally produce a TM peptide that proves challenging to dissolve, even in a detergent solution. Below we provide a detergent reconstitution procedure modified from [12] to include additional pathways for solubilization of more challenging TM peptides.

1. Dissolve the peptide in a minimum volume of HFIP (or TFA; *see* **Note 11**). Cap vial and gently mix (*see* **Note 11**). The peptide should readily dissolve, and the resulting solution clear and free of visible aggregate (*see* **Note 12**).

2. Evaporate HFIP (or TFA) in fume hood at room temperature under a gentle N_2 stream until dry; the peptide will adhere as a film to vessel sides. Care should be taken to avoid blowing the peptide solution out of the vial.

3. Add sufficient TFE to the dried peptide for a final concentration of ~ 1–5 mM. Cap vial and gently mix gently. The peptide should readily dissolve, and the solution should be clear, free of visible aggregate, and remain nonviscous (*see* **Note 13**). If peptides are to be stored in TFE (*see* **Note 14**), blanket the solution with N_2 gas, ensure the screw caps are tightly sealed, and place at 4 °C.

4. Aliquot the desired volume of TFE-dissolved peptide in a clean tube, mix with an appropriate volume of detergent solution in ultrapure water, and add ultrapure water to yield a volume ratio of 16:1 (*see* **Note 15**).

5. Mix the peptide/TFE-detergent-water solution by gentle agitation for 15 min at room temperature.

6. Flash-freeze the above solution in a dry ice/ethanol bath, then lyophilize at or below −70 °C until dry (*see* **Note 16**).

7. Add an appropriate volume of water or aqueous buffer to the dry peptide/detergent mixture (*see* **Note 17**). Gently swirl to mix. Dry components should dissolve immediately and produce a clear solution free of aggregate (*see* **Note 18**).

4 Notes

1. Prediction algorithms have not yet achieved 100 % accuracy in discriminating TM and non-TM residues [13] and generally define TM sequence ends with low precision, with even the best methods in error by four residues at each of the N- and C-terminal ends of helical TM segments [14]. The imprecision associated with TM helix start and end site prediction is illustrated in Fig. 1 with the second TM segment of bovine myelin proteolipid protein (PLP TM2). PLP TM2 was selected as an example because the exclusion of a single residue (Phe 90) at its C-terminus abrogated the ability of the peptide to oligomerize [15]. Selection of appropriate TM sequence termini may be key to the behavior of the isolated peptide in certain cases.

2. Prediction accuracy and reliability may be improved by using an array of methods and a simple "majority rules" approach [13]. For this reason, our group typically uses consensus prediction by a variety of methods to guide the selection of TM peptide sequences (cf. ref. 15). As illustrated in Fig. 1, a general rule of thumb is to incorporate a given residue into the TM peptide sequence if it is identified as TM in at least half of the predictions. We estimate that a minimum of 10–15 predictions should be made, noting that the consensus PLP TM2 peptide sequence derived from 24 predictions (Fig. 1) is identical to the sequence derived from a simple majority prediction from a 13-member subset of these web servers (*see* ref. 15).

3. If several variants of a peptide will be prepared (e.g., various lengths or a series of Gly to Val replacements), it may be worthwhile to calculate the number of Lys residues required based on the peptide sequence with the highest total Liu–Deber hydrophobicity value.

4. A Liu–Deber total hydrophobicity score of <10 does not guarantee water solubility and/or lack of aggregation. Peptides with a net charge of zero tend to be poorly soluble in water, even if their sequences consist largely of hydrophilic residues.

5. The chemistry of the C-terminus is determined by the identity of the linker on the resin used for peptide synthesis; acylation of the peptide N-terminus is performed by "capping" the N-terminal amino group prior to cleavage of the peptide from the resin (*see* ref. 11).

6. The vast majority of helical membrane proteins follow the "positive-inside" rule [16], such that the basic residues Lys and Arg are more abundant on the cytoplasmic side of the membrane. For this reason, we recommend placement of odd Lys residues at the TM peptide terminus expected to reside on the cytoplasmic side of the membrane. Many of the web servers listed in Table 1 predict the topology of membrane-spanning regions.

7. Lys tags may not be optimally suited for all TM peptide applications. Other tags that have proven useful in improving TM peptide solubility include His and *N*-methyl Gly (Sar) [17]. However, Sar-tags may only be effective when used in combination with Lys residues [17]. If possible, Sar should be placed at the N-terminal side of peptide sequences for ease of peptide synthesis. Arg is not recommended as a polar residue tag due to peptide deprotection issues [17].

8. Lys-tag inclusion in some instances may unmask inherent conformational instability and/or plasticity in helix–helix interactions [18] and may affect self-assembly in weakly interacting segments [19, 20]. In these cases, placement of Lys and/or alternate tag residues at the TM peptide terminus distal from the putative site of peptide interaction may be advantageous. For example, a peptide that self-assembles at a site near its N-terminus might be solubilized using a Sar-tag at the peptide N-terminus and a Lys-tag at the C-terminus. Increases in buffer ionic strength (e.g., by adding NaCl) may also help to minimize repulsive electrostatic interactions between the polypositively charged Lys groups at peptide N- and/or C-termini.

9. Cys and Met residues may be subject to oxidation. Methionine sulfoxide can be reduced back to Met by treatment with dithiothreitol (DTT). Peptide sequences containing Cys residues may also require usage of DTT and/or other reducing agent(s) during handling in aqueous buffers to break and/or prevent formation of inter-peptide disulfide linkages (or intra-peptide linkages in sequences with multiple Cys residues). Substitution of Cys and/or Met with an appropriate residue(s) may therefore be preferred in certain cases. Many "Cys-less" membrane proteins retain functionality (e.g., *E. coli* lactose permease [21], cardiac phospholamban [22, 23], human cystic fibrosis transmembrane conductance regulator [24], *E. coli* small multidrug resistance protein [25], and human P-glycoprotein [26]), although activity may be reduced vs. wild-type protein. Cys is most often replaced with Ser or Ala, but in some cases, Val, Leu, or the non-proteinogenic residue α-amino-*n*-butyric acid (Abu), which has an isosteric (CH_2CH_3) side chain, may be utilized, depending on the role(s) of Cys in the native sequence and/or the peptide properties desired by the end user. Norleucine (Nle, side chain $CH_2CH_2CH_3$) may be used to replace Met.

10. Asn or Gln as the N-terminal residue, Pro and/or Gly in the two positions preceding the C-terminus, and Asp–Asn, Asp–Ala, Asp–Gly, Asp–Pro, or Asp–Ser pairs may undergo sequence-dependent side reactions during SPPS and/or peptide cleavage. *See* ref. 11 for details. Distribution of Lys and/or other polar residue tags may be used to avoid undesirable residue identities at the N- and/or C-terminus.

11. Glass syringes should be used to accurately aliquot organic solvents. We recommend testing solubility in HFIP first. It is rare to encounter a TM peptide that is insoluble in HFIP. However, if the TM peptide proves HFIP-insoluble, continue to **Note 2**, and then either repeat **Note 1** with TFA or move to **Note 3**. We caution that certain peptide sequences are unstable in strong acids such as TFA. Note that HFIP is corrosive and highly volatile, and that TFA is a highly corrosive, highly volatile strong acid. Perform all manipulations involving HFIP or TFA with care in a fume hood, using appropriate personal protective equipment.

12. It is rare to encounter a TM peptide that is insoluble in HFIP and in TFA. If this proves to be the case, continue to **Note 2** to evaporate HFIP or TFA, and then proceed with **Note 3**.

13. The TM peptide concentration range given represents a suggested starting point. A given peptide may be more or less soluble in TFE. TFE is corrosive, volatile, and flammable in liquid and vapor form. Perform all manipulations in a fume hood with appropriate personal protective equipment.

14. We maintain certain water-insoluble TM peptides as stock solutions in TFE, stored under N_2 in PTFE-lined, screw-capped glass vials at 4 °C.

15. Screw cap glass vials are not required for this procedure; we typically use conical polypropylene screw cap tubes. The volumes used depend on the stock concentration and the volume and concentration of detergent desired in the final sample. For example, to prepare a 1 mL sample of 0.2 mM TM peptide reconstituted with 1 % (w/v) SDS, one would use 0.1 mL of a 2 mM TM peptide TFE stock solution, mixed with 1 mL of 1 % (w/v) SDS, and diluted with 16 mL of ultrapure water. We do not recommend addition of buffer components at this stage due to the potential for precipitation by TFE. Certain detergents may also be precipitated by high concentrations of TFE; this may be mitigated by dilution of the detergent solution with ultrapure water and addition of a larger volume of less concentrated detergent.

16. Care should be taken to minimize loss of peptide and detergent during lyophilization. This may be accomplished by ensuring that total solution volume does not exceed ~1/4 of the volume of the tube placed in the lyophilizer.

17. In the example above, the dry peptide/SDS mixture would be rehydrated with 1 mL of water or appropriate aqueous buffer.

18. Detergent reconstitution of TM peptides is sensitive to the molar ratio of protein/detergent molecules and to the critical micelle concentration (CMC) of the detergent. It is recommended to maintain detergent concentration above the CMC. Note that detergent CMC may be affected by the presence of high concentrations of protein and/or the ionic strength of the buffer.

Acknowledgements

We wish to thank Derek P. Ng for assistance in preparation of Fig. 1. This work was supported, in part, by grants to C.M.D. from the Canadian Institutes of Health Research (CIHR FRN-5810) and the Natural Science and Engineering Research Council of Canada (NSERC I2I Grant). A.R. was the recipient of a Research Training Centre (RESTRACOMP) award from the Hospital for Sick Children and held a postdoctoral award from the CIHR Strategic Training Program in Protein Folding: Principles and Diseases.

References

1. Overington JP, Al-Lazikani B, Hopkins AL (2006) How many drug targets are there? Nat Rev 5(12):993–996

2. Yildirim MA, Goh KI, Cusick ME, Barabasi AL, Vidal M (2007) Drug-target network. Nat Biotechnol 25(10):1119–1126

3. Bordag N, Keller S (2010) Alpha-helical transmembrane peptides: a "divide and conquer" approach to membrane proteins. Chem Phys Lipids 163(1):1–26

4. Ulmschneider MB, Sansom MS, Di Nola A (2005) Properties of integral membrane protein structures: derivation of an implicit membrane potential. Proteins 59(2):252–265

5. Melnyk RA, Partridge AW, Deber CM (2001) Retention of native-like oligomerization states in transmembrane segment peptides: application to the Escherichia coli aspartate receptor. Biochemistry 40(37):11106–11113

6. Rath A, Melnyk RA, Deber CM (2006) Evidence for assembly of small multidrug resistance proteins by a "two-faced" transmembrane helix. J Biol Chem 281(22):15546–15553

7. Oates J, Hicks M, Dafforn TR, DiMaio D, Dixon AM (2008) In vitro dimerization of the bovine papillomavirus E5 protein transmembrane domain. Biochemistry 47(34):8985–8992

8. Gan SW, Xin L, Torres J (2007) The transmembrane homotrimer of ADAM 1 in model lipid bilayers. Protein Sci 16(2):285–292

9. Torres J, Wang J, Parthasarathy K, Liu DX (2005) The transmembrane oligomers of coronavirus protein E. Biophys J 88(2):1283–1290

10. Tulumello DV, Deber CM (2012) Efficiency of detergents at maintaining membrane protein structures in their biologically relevant forms. Biochim Biophys Acta 1818(5):1351–1358

11. Amblard M, Fehrentz JA, Martinez J, Subra G (2005) Fundamentals of modern peptide synthesis. Methods Mol Biol 298:3–24

12. Duarte AM, Wolfs CJ, van Nuland NA, Harrison MA, Findlay JB, van Mierlo CP, Hemminga MA (2007) Structure and localization of an essential transmembrane segment of the proton translocation channel of yeast H+–V-ATPase. Biochim Biophys Acta 1768(2):218–227

13. Tusnady GE, Simon I (2010) Topology prediction of helical transmembrane proteins: how far have we reached? Curr Protein Pept Sci 11(7):550–561

14. Cuthbertson JM, Doyle DA, Sansom MS (2005) Transmembrane helix prediction: a comparative evaluation and analysis. Protein Eng Des Sel 18(6):295–308

15. Ng DP, Deber CM (2010) Deletion of a terminal residue disrupts oligomerization of a transmembrane alpha-helix. Biochem Cell Biol 88(2):339–345

16. von Heijne G, Gavel Y (1988) Topogenic signals in integral membrane proteins. Eur J Biochem 174(4):671–678

17. Melnyk RA, Partridge AW, Yip J, Wu Y, Goto NK, Deber CM (2003) Polar residue tagging of transmembrane peptides. Biopolymers 71(6):675–685

18. Parthasarathy K, Ng L, Lin X, Liu DX, Pervushin K, Gong X, Torres J (2008) Structural flexibility of the pentameric SARS coronavirus envelope protein ion channel. Biophys J 95(6):L39–L41

19. Lew S, Caputo GA, London E (2003) The effect of interactions involving ionizable residues flanking membrane-inserted hydrophobic helices upon helix-helix interaction. Biochemistry 42(36):10833–10842

20. Iwamoto T, You M, Li E, Spangler J, Tomich JM, Hristova K (2005) Synthesis and initial characterization of FGFR3 transmembrane domain: consequences of sequence modifications. Biochim Biophys Acta 1668(2):240–247

21. van Iwaarden PR, Pastore JC, Konings WN, Kaback HR (1991) Construction of a functional lactose permease devoid of cysteine residues. Biochemistry 30(40):9595–9600

22. Karim CB, Paterlini MG, Reddy LG, Hunter GW, Barany G, Thomas DD (2001) Role of cysteine residues in structural stability and function of a transmembrane helix bundle. J Biol Chem 276(42):38814–38819

23. Afara MR, Trieber CA, Glaves JP, Young HS (2006) Rational design of peptide inhibitors of the sarcoplasmic reticulum calcium pump. Biochemistry 45(28):8617–8627

24. Cui L, Aleksandrov L, Hou YX, Gentzsch M, Chen JH, Riordan JR, Aleksandrov AA (2006) The role of cystic fibrosis transmembrane conductance regulator phenylalanine 508 side chain in ion channel gating. J Physiol 572(Pt 2):347–358

25. Mordoch SS, Granot D, Lebendiker M, Schuldiner S (1999) Scanning cysteine accessibility of EmrE, an H+–coupled multidrug transporter from Escherichia coli, reveals a hydrophobic pathway for solutes. J Biol Chem 274(27):19480–19486

26. Loo TW, Clarke DM (1995) Membrane topology of a cysteine-less mutant of human P-glycoprotein. J Biol Chem 270(2):843–848

27. Greer JM, Lees MB (2002) Myelin proteolipid protein–the first 50 years. Int J Biochem Cell Biol 34(3):211–215

28. Chen CP, Rost B (2002) State-of-the-art in membrane protein prediction. Appl Bioinformatics 1(1):21–35

29. Punta M, Forrest LR, Bigelow H, Kernytsky A, Liu J, Rost B (2007) Membrane protein prediction methods. Methods 41(4):460–474

30. Hennerdal A, Elofsson A (2011) Rapid membrane protein topology prediction. Bioinformatics 27(9):1322–1323

31. Liu LP, Deber CM (1998) Guidelines for membrane protein engineering derived from de novo designed model peptides. Biopolymers 47(1):41–62

32. Cserzo M, Eisenhaber F, Eisenhaber B, Simon I (2002) On filtering false positive transmembrane protein predictions. Protein Eng 15(9):745–752

33. Tusnady GE, Simon I (2001) The HMMTOP transmembrane topology prediction server. Bioinformatics 17(9):849–850

34. Shen H, Chou JJ (2008) MemBrain: improving the accuracy of predicting transmembrane helices. PLoS One 3(6):e2399

35. Pierleoni A, Indio V, Savojardo C, Fariselli P, Martelli PL, Casadio R (2011) MemPype: a pipeline for the annotation of eukaryotic membrane proteins. Nucleic Acids Res 39(Web Server issue):W375–W380

36. Jones DT (2007) Improving the accuracy of transmembrane protein topology prediction using evolutionary information. Bioinformatics 23(5):538–544

37. Nugent T, Jones DT (2009) Transmembrane protein topology prediction using support vector machines. BMC Bioinforma 10:159

38. Snider C, Jayasinghe S, Hristova K, White SH (2009) MPEx: a tool for exploring membrane proteins. Protein Sci 18(12):2624–2628

39. Viklund H, Elofsson A (2008) OCTOPUS: improving topology prediction by two-track ANN-based preference scores and an extended topological grammar. Bioinformatics 24(15):1662–1668

40. Liakopoulos TD, Pasquier C, Hamodrakas SJ (2001) A novel tool for the prediction of transmembrane protein topology based on a statistical analysis of the SwissProt database: the OrienTM algorithm. Protein Eng 14(6):387–390

41. Rost B, Casadio R, Fariselli P, Sander C (1995) Transmembrane helices predicted at 95% accuracy. Protein Sci 4(3):521–533

42. Reynolds SM, Kall L, Riffle ME, Bilmes JA, Noble WS (2008) Transmembrane topology and signal peptide prediction using dynamic bayesian networks. PLoS Comput Biol 4(11):e1000213

43. Kall L, Krogh A, Sonnhammer EL (2004) A combined transmembrane topology and signal peptide prediction method. J Mol Biol 338(5):1027–1036

44. Viklund H, Elofsson A (2004) Best alpha-helical transmembrane protein topology predictions are achieved using hidden Markov models and evolutionary information. Protein Sci 13(7):1908–1917

45. Bernsel A, Viklund H, Falk J, Lindahl E, von Heijne G, Elofsson A (2008) Prediction of membrane-protein topology from first principles. Proc Natl Acad Sci USA 105(20): 7177–7181

46. Hirokawa T, Boon-Chieng S, Mitaku S (1998) SOSUI: classification and secondary structure prediction system for membrane proteins. Bioinformatics 14(4):378–379

47. Juretic D, Zoranic L, Zucic D (2002) Basic charge clusters and predictions of membrane protein topology. J Chem Inf Comput Sci 42(3):620–632

48. Viklund H, Bernsel A, Skwark M, Elofsson A (2008) SPOCTOPUS: a combined predictor of signal peptides and membrane protein topology. Bioinformatics 24(24):2928–2929

49. Lo A, Chiu HS, Sung TY, Lyu PC, Hsu WL (2008) Enhanced membrane protein topology prediction using a hierarchical classification method and a new scoring function. J Proteome Res 7(2):487–496

50. Deber CM, Wang C, Liu LP, Prior AS, Agrawal S, Muskat BL, Cuticchia AJ (2001) TM Finder: a prediction program for transmembrane protein segments using a combination of hydrophobicity and nonpolar phase helicity scales. Protein Sci 10(1):212–219

51. Krogh A, Larsson B, von Heijne G, Sonnhammer EL (2001) Predicting transmembrane protein topology with a hidden Markov model: application to complete genomes. J Mol Biol 305(3):567–580

52. Ganapathiraju M, Balakrishnan N, Reddy R, Klein-Seetharaman J (2008) Transmembrane helix prediction using amino acid property features and latent semantic analysis. BMC Bioinforma 9(Suppl 1):S4

53. von Heijne G (1992) Membrane protein structure prediction. Hydrophobicity analysis and the positive-inside rule. J Mol Biol 225(2): 487–494

Chapter 12

Engineering and Utilization of Reporter Cell Lines for Cell-Based Assays of Transmembrane Receptors

Matthew W. Lluis and Hang Yin

Abstract

Transmembrane receptors, a subset of integral membrane proteins, are the receivers that transduce an extracellular chemical message into an intracellular response. Accordingly, these proteins are of particular interest in the scientific community and are probably best studied as part of a cellular system. Herein, we detail the engineering of a fluorescent and bioluminescent reporter cell line for a transmembrane receptor and how to employ it in a directed evolution screen that identifies peptide regulators of receptor activity.

Key words Transmembrane receptors, Green-fluorescent protein, Luciferase, Reporter cell line, Directed evolution, Traptamers

1 Introduction

Transmembrane receptors receive chemical stimuli from the extracellular milieu and propagate them as intracellular signals [1]. Therefore, they represent a vital component of a cell's sensory system. As a result, the biochemistry of these proteins is of high interest in the scientific community [1–4]. However, due to the high level of difficulty in recombinantly expressing integral membrane proteins, the in vitro study of them is challenging [5]. An alternative, and perhaps more biologically relevant, method of studying transmembrane receptors is to develop a cell-based assay by which the protein's biochemical properties can be investigated. Unlike many water-soluble proteins that function in isolation by simply converting a substrate to product, transmembrane receptors function in integrated signaling cascades that involve several other factors [6, 7]. Therefore, cell-based assays allow for the researcher to probe how changes in the chemistry of the receptor affect a signaling cascade at the cellular level. Additionally, a cell-based assay that uses a fluorescent reporter cell line allows for the utilization of flow cytometry, a powerful and versatile high throughput method [8]. For certain transmembrane receptors, reporter cell lines for

Giovanna Ghirlanda and Alessandro Senes (eds.), *Membrane Proteins: Folding, Association, and Design*,
Methods in Molecular Biology, vol. 1063, DOI 10.1007/978-1-62703-583-5_12, © Springer Science+Business Media, LLC 2013

cell-based assays are commercially available. However, these are almost always plasmid transfected, enzyme reporter cell lines that need to be cultured in selective media and cannot be easily employed with flow cytometry. Therefore, engineering a cell line in house is a cost effective way by which a researcher has the freedom to tailor a cell-based assay to their specific needs.

The following protocol describes how to engineer a fluorescent and luminescent reporter cell line to be used for cell-based assays of signaling pathways, more specifically, transmembrane receptors. Although there are several ways by which this cell line can be used, the most practical is employing it in a directed evolution screen to select short peptide inhibitors or activators of transmembrane receptor signaling. Directed evolution is a powerful method in molecular biology by which proteins that have specific phenotypic properties are selected from a large and diverse library of protein sequences (>10^6 sequences) [9]. The directed evolution method outlined has tremendous potential as it has been successfully used to engineer peptides that activate transmembrane receptors by specifically targeting the receptor's transmembrane domain. These short peptide sequences have been designated "traptamers" as shorthand for transmembrane aptamers [10]. This is a key point as there is a general lack of molecular tools that can probe the functions of transmembrane domains. Furthermore, mounting scientific evidence suggests that many transmembrane receptors, including the G-protein coupled receptors (GPCRs) and the Toll-like receptors, oligomerize through their transmembrane domains and that these interactions are crucial for function [11–18]. DiMaio and coworkers have successfully applied directed evolution to evolve peptides that specifically target the transmembrane domains of platelet-derived growth factor β receptor tyrosine kinase (PDGFβR) and the human erythropoietin receptor (hEPOR) [10, 19]. For these directed evolution screens, peptide sequences were selected from two randomized libraries constructed from the scaffold of the E5 protein of bovine papillomavirus, an oncogenic, 44-amino acid single pass transmembrane helix that naturally binds to the transmembrane domain of PDGFβR [10, 19]. It was previously demonstrated that four amino acids in the E5 sequence are required for binding to PDGFβR (Gln17, Asp33, Cys37, and Cys39) [20–25]. A traptamer library was constructed by randomizing residues 14–30 and 33 of E5 and employed in a cell-based directed evolution screen to select traptamer sequences that specifically activate the PDGFβR [19]. Interestingly, the top-ranked traptamer sequence has virtually no sequence homology to E5 and does not contain any of the four amino acids demonstrated to be required for E5 function [19]. To further investigate whether traptamers that target other transmembrane receptors could be identified by this method, the authors applied this cell-based directed evolution technique to the hEPOR [10]. A ~500,000 sequence traptamer library was constructed by randomizing residues 12–30 (the E5 transmembrane domain) and 33. Specifically, residues 12–30 were randomized in such a manner as to favor the coding of hydrophilic residues, whereas residue 33 was randomized with a hydrophilic amino acid bias [10, 26]. Using the same cell-based directed evolution method, the DiMaio group converged the ~500,000 sequence traptamer library down to a consensus transmembrane sequence of ILVGTLIVLIPVLIVLVFLYWQ [10]. This sequence is not homologous to the E5 transmembrane domain (LVAAMQLLLLLFLLLLFFLVYWD), and the selected traptamer's specificity for the hEPOR was cross-validated by biophysical methods [10]. These studies are representative of the practicality of applying cell-based directed evolution to transmembrane receptor targets.

As stated previously, perhaps the most practical use of the reporter cell line is for directed evolution of transmembrane receptor targeting peptides that can serve as inhibitors or activators of receptor activity. Subheading 3.2 was adopted from refs. 10 and 26 and describes how to construct a recombinant retrovirus population that will stably transduce a peptide library into a reporter cell population for a targeted transmembrane receptor [10, 26] (Fig. 1). Once integrated into the reporter cell genomic DNA, each peptide sequence will be overexpressed from the viral promoter of the 5′ long terminal repeat (5′ LTR). Using flow cytometry under activated or inactivated receptor conditions, cells that have been transduced with peptide sequences that inhibit or enhance transmembrane receptor activity can be identified and enriched in the population, whereas cells that have been transduced with inert peptides can be removed [8]. Reiterative cycles of selection by flow cytometry will greatly reduce the size and diversity of the original peptide library allowing for identification of peptide sequences that produce the desired phenotype of receptor inhibition or activation in the reporter cell line. In order to utilize this protocol, a fluorescent reporter cell line for a target transmembrane receptor must be available or constructed following Subheading 3.1. This protocol can also be simplified for low throughput screening of peptide inhibitors/activators by only transducing a handful of known peptide sequences as opposed to a large peptide library. Additionally, because the pGreenfire™ system used in Subheading 3.1 expresses luciferase in addition to GFP, the investigation of peptide inhibitors/activators of transmembrane receptors can also be performed using a bioluminescent assay. However, bioluminescence cannot be easily incorporated into a directed evolution protocol. For those wanting to exercise directed evolution, a peptide library inserted into a retroviral expression vector must be available or constructed. Furthermore, because the target is a membrane embedded receptor, researchers constructing a library will have to make special considerations for the library design. For this one is referred to refs. 10 and 26.

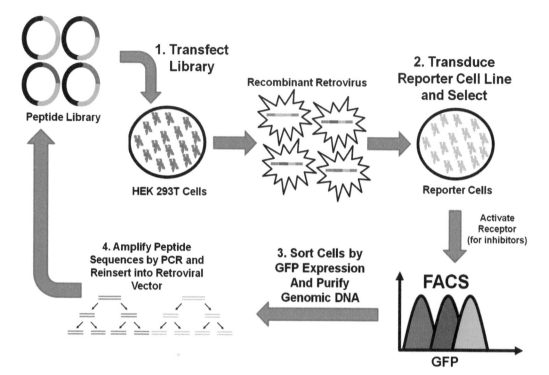

Fig. 1 The directed evolution scheme for identification of peptide inhibitors of activators of transmembrane receptors. A peptide library encoding different peptide sequences (represented as *green, purple, blue,* and *red*) and antibiotic resistance (*light blue*) is transiently transfected into HEK 293T cells (virus packaging cell line). Recombinant virus is then recovered and used to transduce a reporter cell line population. With the target receptor under activated (identifying inhibitors) or inactivated (identifying activators) conditions, the transduced reporter cell population is sorted by flow cytometry. The genomic DNA of the selected cells is recovered, and the encoded library sequences are amplified by PCR and inserted back into the original retroviral vector to construct a second generation library

2   Materials

Prepare all solutions using ultrapure, deionized water. Only use analytical grade reagents of the highest quality. All solutions to be used with mammalian cell culture have to be sterile filtered through a 0.22 mM filter into a sterile container before use.

2.1   Tissue Culture Growth Medium and Solutions

1. DMEM-10+PS (growth media for Human Embryonic Kidney 293T cells): Add 100 mL of fetal bovine serum (FBS) and 10 mL of 100× penicillin–streptomycin to 890 mL of Dulbecco's Modified Eagle Medium. Final concentration is 10 % FBS, 100 IU penicillin, and 100 μg/mL streptomycin. Store at 4 °C. Growth media is always warmed to 37 °C before use.

2. Opti-MEM® I (Life Technologies, www.lifetechnologies.com): reduced serum media.

3. DMEM-HEPES: 25 mM HEPES added to DMEM-10-PS.

4. Phosphate Buffered Saline (PBS): Dissolve 4 g of NaCl, 0.10 g of KCl, 1.39 g of $Na_2HPO_4.2H_2O$, and $0.135 \times g$ of KH_2PO_4 in 400 mL of H_2O. Add HCl until a pH of 7.4. Adjust the volume to 500 mL with H_2O. Store at room temperature.

5. Trypsin-EDTA: 0.25 % w/v Trypsin-0.53 mM EDTA solution. Store at 4 °C.

2.2 Transfection and Transduction Reagents

1. pGreenfire™ Lentiviral Plasmid (Systems Biosciences, Mountain View, CA, www.systembio.com): Lentiviral fluorescent reporter vector for human immunodeficiency virus-1 (HIV-1) located downstream of the transcriptional response element (TRE) of the signaling pathway of interest (Fig. 2, see Notes 1 and 2).

2. pPACKH1™ Packaging Plasmids (Systems Biosciences, Mountain View, CA, www.systembio.com): HIV-1 packaging plasmids (see Note 1).

3. pMSCVpuro Retroviral Vector (Clontech Laboratories, www.clontech.com): Murine Stem Cell Virus Retroviral Vector with Puromycin resistance.

4. pCL-Eco (IMGENEX, www.imgenex.com): Retroviral packaging vector.

5. pVSV-G (Clontech Laboratories, www.clontech.com): Retroviral pseudotyping vector.

6. Polyethyleneimine (PEI): Dissolve PEI in ultrapure water. Using HCl, adjust the pH to 7.0. Add water to a final concentration of 1 mg/mL. Filter sterilize through a 0.22 μM filter. Store at −80 °C (see Note 3).

7. Polybrene: Dissolve polybrene in ultrapure water. Add water to a final concentration of 4 mg/mL. Filter sterilize through a 0.22 μM filter. Store at −80 °C.

8. Retro-Concentin™ (5× solution, Systems Biosciences, Mountain View, CA, www.systembio.com): Retrovirus concentrating reagent.

2.3 Mammalian Cell Lines

1. Human Embryonic Kidney 293T (HEK 293T) cells (see Note 4).

2. Target cell line (see Note 5).

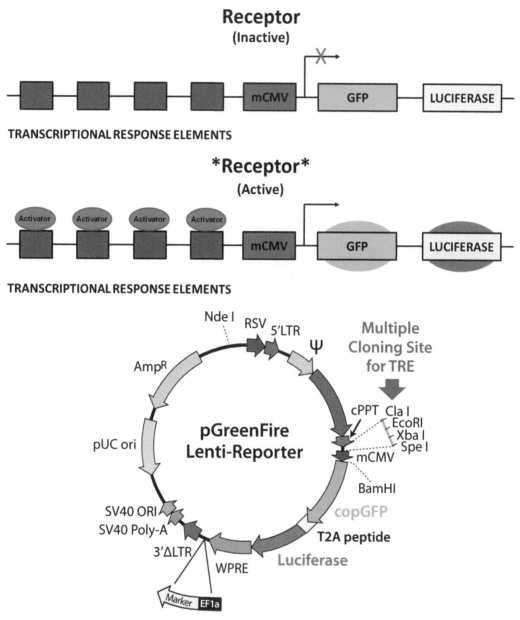

Fig. 2 The pGreenfire™ reporter system. (*Top*) GFP and luciferase are located downstream of a minimal cyto-megalovirus promoter (mCMV). The mCMV promoter is transcriptionally dormant when the target transmembrane receptor is in its respective inactive state. Activation of a signaling pathway by a receptor results in specific transcriptional activators (*blue ovals*) binding to transcriptional response elements (TREs, *blue squares*) upstream of the mCMV, which amplifies expression of GFP and luciferase. (*Bottom*) The general vector map of a pGreenfire lentiviral reporter vector. The genomic elements flanked by the 5′ and 3′ LTRs will be stably inserted into the reporter cell line genomic DNA. The selection marker is available as Puromycin or Neomycin resistance. Image reprinted with permission from Systems Biosciences (SBI) (www.systembio.com)

3 Methods

3.1 Engineering a Fluorescent/ Luminescent Reporter Cell Line

All procedures are carried out at room temperature in a viral containment laminar flow hoods following biosafety level 2 (BSL-2) safety protocols.

Culturing of HEK293T Cells

1. Thaw a frozen stock of HEK 293T cells by placing the cells in a 37 °C water bath.

2. Once the cell stock has thawed, add all of the cells to a T175 tissue culture flask (*see* **Note 6**).

3. Add 20 mL of growth medium dropwise (*see* **Note 7**).

4. Grow cells in a tissue culture incubator at 37 °C until 80–90 % confluency is reached.

Day 1: Transfection of Virus Packaging Cells

1. Aspirate off DMEM-10 + PS from an 80 % to 90 % confluent T175 flask culturing HEK 293T cells.

2. Add 5 mL of sterile PBS (*see* **Note 8**).

3. Aspirate off PBS and add 5 mL of trypsin-EDTA. Allow the trypsin to digest for 5 min (*see* **Note 9**).

4. Add 25 mL of DMEM-10 + PS to the digestion and transfer the cellular suspension to a sterile 50 mL conical tube.

5. Spin down the cellular suspension for 10 min at $500 \times g$.

6. Aspirate off all media and resuspend cells in fresh DMEM-10 + PS. Calculate the cellular density using a hemacytometer.

7. In a sterile, 15 mL falcon tube, mix 1.5 mL of Opti-MEM® I, 90 μL of 1 mg/mL PEI, and a total of 15 μg of DNA. The 15 μg of DNA will be divided into eight 2/3 μg of packaging plasmids (pPACKH1-GAG and pPACKH1-REV), 4 1/3 μg lentiviral plasmid (pGreenfire™), and 2 μg of pseudotyping plasmid (pVSV-g) (*see* **Note 10**).

8. Incubate the transfection mixture at 37 °C for 20 min.

9. In a 10 cm culture dish, plate out a total of 5×10^6 HEK 293T cells in 10 mL of DMEM-10 + PS (*see* **Note 11**).

10. Immediately after plating HEK 293T cells, add the transfection mixture dropwise to the cell containing media.

11. Swirl the dish gently to evenly mix the transfection solution.

12. Place the cells in a tissue culture incubator at 37 °C for 18 h.

13. In a 15 cm petri dish, plate out 3×10^6 of target cells.

Day 2: Passaging of Virus Packaging Cells

14. Aspirate off media and wash the cells with 5 mL of PBS.

15. Aspirate off the PBS and add 5 mL of trypsin-EDTA. Allow the trypsin to digest for 5 min.

16. Add 10 mL of DMEM-10+PS and transfer the cellular suspension to a 15 mL conical tube.

17. Spin down the cellular suspension for 10 min at $500 \times g$.

18. Aspirate off all media and resuspend cells in 20 mL of DMEM-10+PS.

19. Plate out all of the cellular suspension on a 15 cm petri dish.

Day 3: First Transduction of Target Cells

20. After 24 h (48 h post transfection), remove the virus containing DMEM-10-PS media from the transfected 293T cells with a pipet and filter this media through a 0.22 μm filter (*see* **Note 12**).

21. Add 20 mL of fresh DMEM-10-PS to the HEK 293T cells and place them in a tissue culture incubator at 37 °C.

22. Centrifuge target cells in a sterile 50 mL conical tube at $500 \times g$ for 10 min.

23. Aspirate media from target cells and replace with filtered virus containing DMEM-10+PS media.

24. Add polybrene to a final concentration of 4 μg/mL and replate out target cells in 15 cm petri dish (*see* **Note 13**).

25. Place these cells in a tissue culture incubator at 37 °C.

Day 4: Second Transduction of Target Cells

26. After 24 h (72 h post transfection), remove the virus containing DMEM-10+PS media from the transfected 293T cells with a pipet and filter this media through a 0.22 μm filter.

27. In a sterile, 50 mL conical tube, add the once transduced target cell culture and centrifuge at $500 \times g$ for 10 min.

28. Aspirate the media from target cells and replace with filtered virus containing DMEM-10+PS media.

29. Add polybrene to a final concentration of 4 μg/mL and replate out target cells in 15 cm petri dish.

30. Place these cells in a tissue culture incubator at 37 °C.

Day 5: Exchange of Media to Remove Virus

31. In a 50 mL conical tube, add the twice transduced target cell culture and centrifuge at $500 \times g$ for 10 min.

32. Aspirate off all of the virus containing media and resuspend in 20 mL of fresh growth medium (*see* **Note 14**).

Day 7: Selection of Target Cells

33. In parallel, add selective antibiotic to the lentivirally transduced target cell culture and an untransduced target cell culture (*see* **Note 15**).

Recovery of Transduced Cells by Flow Cytometry

34. Once antibiotic selection is complete, replate transduced target cells in fresh growth medium that is free of the selection antibiotic. Allow the selected cells to recover for 24–48 h.

35. Split the transduced target cell culture in half. Chemically treat one half of this target cell culture to activate the target transmembrane receptor.

36. Sort the transduced target cells under activated receptor conditions monitoring GFP expression (Fig. 3). Use an untransduced, unactivated receptor target cell culture and the unactivated, transduced target cells as controls (*see* **Note 16**).

37. Continue culturing selected cells for use in cell-based assays or flash freeze aliquots of selected cells for storage at –80 °C.

38. Repeat **step 35** until the receptor response is optimized.

3.2 Cell-Based Directed Evolution Screen for Peptide Inhibitors of Transmembrane Receptors

All procedures are carried out at room temperature in a viral containment laminar flow hoods under BSL-2 safety protocols.

Day 1: Transfection of Virus Packaging Cell Line

1. Plate out 5×10^6 HEK 293T cells in 4, 10 cm plates with 10 mL of DMEM-10 + PS (as detailed in Subheading 3.1).

2. In a 15 mL conical tube, mix 1 mL of Opti-MEM® I, 60 µL of PEI, 2 µg of pVSV-G, 3 µg of pCL-Eco, and 5 µg of peptide library (or single peptide sequence) in pMSCVpuro. This is the peptide library transfection. In another 15 mL conical tube, mix 1 mL of Opti-MEM® I, 60 µL of PEI, 2 µg of pVSV-G, 3 µg of pCL-Eco, and 5 µg of empty pMSCVpuro (an open reading frame has not been inserted into the multiple cloning site and consequently cannot express a peptide). This is the control transfection. Incubate these transfection mixtures at 37 °C for 20 min.

3. Using a pipetman, split each of the transfections evenly onto two HEK 293T cultures in 10 cm petri dishes and place in a tissue culture incubator at 37 °C for 12 h.

Day 2: Passaging of Virus Packaging Cell Line

4. Passage each 10 cm dish of 293T cells into a 15 cm dish (as detailed in Subheading 3.1) and place in a tissue culture incubator at 37 °C for 24 h.

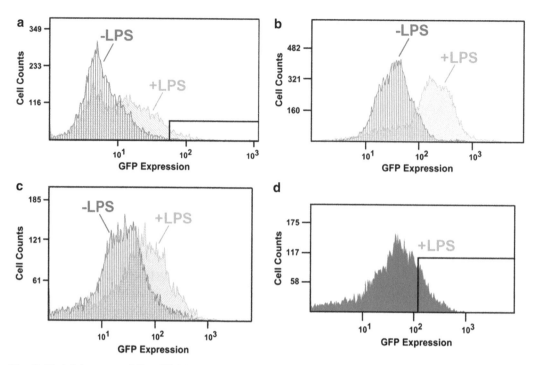

Fig. 3 (*Top*) A human cell line (U937) that endogenously expresses an integral membrane protein, Toll-like receptor 4 (TLR4), was transduced with a pGreenfire lentiviral reporter vector with an NF-κB response element. (**a**) The first round of sorting. Lipopolysaccharide (LPS, TLR4 agonist) was added to the transduced cell culture at a concentration of 1 μg/mL. The *red* histogram represents the untransduced cell population. The *green* histogram represents the transduced and activated cell population. Only the top 5 % of cells expressing GFP were collected (represented by the *bar region*). (**b**) The fluorescence response of the reporter cell line. After two rounds of sorting for the top 5 % of cells expressing GFP, the resulting reporter cell population was cultured and tested for its response to LPS. The *red* population represents an inactivated population (−LPS), and the *green* population represents an activated population (+LPS). (*Bottom*) Directed evolution screening for peptide inhibitors of TLR4 activation. (**c**) Histograms of control reporter cell populations. A reporter cell population was transduced with retrovirus lacking an open reading frame. One half of this cell population was treated with 1 μg/mL LPS to activate the TLR4 receptor. The inactivated population is shown in *red*. The activated LPS-treated population is shown in *green*. (**d**) The peptide library expressing reporter cell population. A reporter cell population was transduced with retrovirus expressing a peptide library and treated with 1 μg/mL LPS. The top 15 % of cells expressing GFP (represented as the *bar region*) were discarded. The remaining cells were recovered and used to construct a subsequent generation peptide library

Day 3: Harvesting of Virus

5. Using a pipet, remove media from peptide library transfected 293T cells and filter through a 0.22 μM filter. Combine filtered media and transfer to a 50 mL conical tube and add Retro-Concentin® to a final concentration of 1× (*see* **Note 17**). Repeat for the control transfected media.

6. Incubate Retro-Concentin®/viral media mixture overnight at 4 °C.

Day 4: Transduction of Reporter Cell Line

7. Spin down viral media for 30 min at $1,500 \times g$ (*see* **Note 18**).

8. Aspirate off media ensuring not to disturb the viral pellet.

9. Spin down viral media for 5 min at $1,500 \times g$.

10. Aspirate off any residual media and resuspend each viral pellet in 500 mL of DMEM-HEPES.

11. In a 12 well tissue culture plate, add 500 μL of a 1×10^6 cells/mL reporter cell culture to three of the wells.

12. Add 500 μL of resuspended library virus to one of reporter cell cultures and 500 μL of resuspended control virus to another reporter cell culture. To the third culture, add only DMEM-HEPES (*see* **Note 19**). Add polybrene to each well to a final concentration of 4 μg/mL.

13. Incubate in a tissue culture incubator at 37 °C for 3 h.

14. Transfer each of the three cultures to a corresponding T25 culture flask. Add 9 mL of reporter cell line growth media and adjust the concentration of polybrene to 4 μg/mL.

Day 6: Selection of Cells

15. To each reporter cell culture, add Puromycin to a final concentration of 1 μg/mL.

Screening of Peptide Library

16. Once antibiotic selection is complete, replate the peptide library expressing pool of cells and the control pools of cells in fresh growth medium free of antibiotic. Allow the cells to recover for 24–48 h.

17. Split the control pool of cells in half. Chemically treat one half of the control cell culture to activate the target transmembrane receptor.

18. For identifying inhibitors of receptor activation: Chemically treat the peptide library expressing cells to activate the target transmembrane receptor.

19. Using a flow cytometer, obtain histograms of the GFP expression (fluorescence) for the activated and inactivated control cell population.

20. Using the control histograms, sort the peptide library expressing cells accordingly for cells that are displaying a reduced fluorescence response or cells that are displaying an increased fluorescence response (*see* **Note 20**).

21. Purify genomic DNA from selected cells. Amplify peptide insert by PCR and religate PCR product into retroviral expression vector (pMSCVpuro) to yield a second generation peptide library with reduced size and diversity.

22. Repeat **steps 1–21** with second generation peptide library and all subsequent generation peptide libraries until library convergence is reached (*see* **Note 21**).

4 Notes

1. This protocol is for the stable insertion of a fluorescent/luminescent reporter DNA cassette into a mammalian cell line via a lentiviral transduction mechanism. Therefore, a successful transduction can only occur if the viral packaging genes are compatible with the LTRs of the lentiviral vector used. This protocol is written for employing Systems Biosciences (SBI) HIV-1 based system for viral transductions, which uses pGreenfire™ lentiviral reporter vectors and pPACKH1™ packaging plasmids.

2. Systems Biosciences (SBI) has a large set of pGreenfire™ lentiviral vectors that detect activation of different signaling pathways by having different TREs located upstream of GFP and luciferase reporter genes. Additionally, each pGreenfire vector is available with a Neomycin or Puromycin selection cassette.

3. When thawed at room temperature, the PEI solution can become cloudy. Before using in a transfection, the PEI must be pre warmed a 50 °C until the solution has become clear.

4. Human Embryonic Kidney 293T cells can be purchased from the American Type Culture Collection (ATCC, www.atcc.org).

5. The target cell line is the mammalian cell line that is to be transduced to yield a reporter cell line. Therefore, the signaling pathway of interest needs to be endogenously or recombinantly expressed in the target cell line.

6. HEK 293T cells should be grown as an adherent culture.

7. In order to avoid thermal shock, media is added dropwise to frozen cell cultures.

8. It is best to add the PBS dropwise as adding it too fast can lift the cells from the surface.

9. It may be necessary to agitate the flask during trypsin digestion in order to completely lift the cells from the surface.

10. Using the pGreenfire™ system, the packaging plasmids will be pPACKH1-GAG and pPACKH1-REV. The lentiviral plasmid will be a pGreenfire™ vector with a TRE that corresponds to the pathway to be monitored. The pseudotyping vector will be pVSV-g. Alternative packaging vectors can be used provided those vectors encode the *gag*, *pol*, *env*, and *rev* genes of HIV. An alternative lentiviral vector can be used provided that vector has LTRs that are compatible with the HIV-1 packaging genes.

11. Swirl the cellular suspension to evenly plate the cells on the 10 cm petri dish.

12. A 0.45 µM filter can also be used.

13. Polybrene is a chemical reagent that increases the efficiency of retroviral transduction; however, it can be toxic to some cells. Therefore, prior to viral transduction, one should determine a concentration of polybrene that is safe to use with their respective target cell type. A polybrene concentration of 4 µg/mL is generally tolerable with most cell types.

14. DMEM is the base media used for HEK293T cells and by default is used for this viral transduction protocol. However, the target cell line may use a different base media and this respective media should be used once the viral transduction steps are completed.

15. This step removes any untransduced cells from the culture prior to cell sorting. Transduced target cells will be resistant to the antibiotic whereas the antibiotic will remain cytotoxic to untransduced cells. Therefore, the untransduced culture is used as the reference point by which antibiotic selection is determined to be complete. This usually takes 2–3 days.

16. Because basal levels of GFP expression will occur even in the absence of pathway activation, the untransduced, inactivated target cell culture allows for sorting of transduced cells from untransduced cells. The difference in GFP expression between the transduced, inactivated target cells (negative control) and the transduced, activated target cells (positive control) represents how easily receptor activation can be detected in this cell line. This difference should be optimized to be as large as possible. Under activated receptor conditions, collecting the most fluorescent cells is most likely to result in collecting the cells that are most responsive to activation (biggest increase in GFP expression).

17. The addition of NaCl (final concentration of 0.4 M NaCl) and polyethylene glycol 6000 (final concentration of 8.5 %) can substitute for Retro-Concentin®.

18. The viral particles should be visible as a gray pellet.

19. The reporter cell culture to which only DMEM-HEPES is added will not be resistant to antibiotic and therefore will serve as the time marker for when antibiotic selection is complete.

20. The activated and inactivated control cell populations will serve as the positive and negative controls of the fluorescence signal. These cell profiles are needed to set the parameters by which the peptide library expressing population will be sorted.

21. Convergence is the point at which fluorescence between the activated control population and the activated library expressing population cannot be increased.

Acknowledgements

We thank the National Science Foundation (CHE 0954819) and the National Institutes of Health (GM 103843 and GM 101279) for financial supports.

References

1. von Heijne G (2007) The membrane protein universe: what's out there and why bother? J Intern Med 261(6):543–557

2. Lappano R, Maggiolini M (2011) G protein-coupled receptors: novel targets for drug discovery in cancer. Nat Rev Drug Discov 10(1): 47–60

3. Overington JP, Al-Lazikani B, Hopkins AL (2006) How many drug targets are there? Nat Rev Drug Discov 5(12):993–996

4. Russ AP, Lampel S (2005) The druggable genome: an update. Drug Discov Today 10(23–24):1607–1610

5. Katzen F, Peterson TC, Kudlicki W (2009) Membrane protein expression: no cells required. Trends Biotechnol 27(8):455–460

6. Maurice P et al (2011) GPCR-interacting proteins, major players of GPCR function. Adv Pharmacol 62:349–380

7. Akira S (2003) Toll-like receptor signaling. J Biol Chem 278(40):38105–38108

8. Daugherty PS, Iverson BL, Georgiou G (2000) Flow cytometric screening of cell-based libraries. J Immunol Methods 243(1–2): 211–227

9. Arnold FH (1998) Design by directed evolution. Accounts Chem Res 31:125–131

10. Cammett TJ et al (2010) Construction and genetic selection of small transmembrane proteins that activate the human erythropoietin receptor. Proc Natl Acad Sci USA 107(8): 3447–3452

11. Caputo GA et al (2008) Computationally designed peptide inhibitors of protein-protein interactions in membranes. Biochemistry 47(33):8600–8606

12. Chin CN, Sachs JN, Engelman DM (2005) Transmembrane homodimerization of receptor-like protein tyrosine phosphatases. FEBS Lett 579(17):3855–3858

13. Finger C, Escher C, Schneider D (2009) The single transmembrane domains of human receptor tyrosine kinases encode self-interactions. Sci Signal 2(89):ra56

14. Li R et al (2004) Dimerization of the transmembrane domain of Integrin alphaIIb subunit in cell membranes. J Biol Chem 279(25): 26666–26673

15. Nemoto W, Toh H (2006) Membrane interactive alpha-helices in GPCRs as a novel drug target. Curr Protein Pept Sci 7(6): 561–575

16. Yin H et al (2007) Computational design of peptides that target transmembrane helices. Science 315(5820):1817–1822

17. Zhang H et al (2002) Integrin-nucleated Toll-like receptor (TLR) dimerization reveals subcellular targeting of TLRs and distinct mechanisms of TLR4 activation and signaling. FEBS Lett 532(1–2):171–176

18. Zhu H et al (2010) Specificity for homooligomer versus heterooligomer formation in integrin transmembrane helices. J Mol Biol 401(5): 882–891

19. Talbert-Slagle K et al (2009) Artificial transmembrane oncoproteins smaller than the bovine papillomavirus E5 protein redefine sequence requirements for activation of the platelet-derived growth factor beta receptor. J Virol 83(19):9773–9785

20. Horwitz BH et al (1988) 44-amino-acid E5 transforming protein of bovine papillomavirus requires a hydrophobic core and specific carboxyl-terminal amino acids. Mol Cell Biol 8(10):4071–4078

21. Klein O et al (1999) The bovine papillomavirus E5 protein requires a juxtamembrane negative charge for activation of the platelet-derived growth factor beta receptor and transformation of C127 cells. J Virol 73(4):3264–3272

22. Klein O et al (1998) Role of glutamine 17 of the bovine papillomavirus E5 protein in platelet-derived growth factor beta receptor activation and cell transformation. J Virol 72(11):8921–8932

23. Schlegel R et al (1986) The E5 transforming gene of bovine papillomavirus encodes a small, hydrophobic polypeptide. Science 233(4762): 464–467

24. Sparkowski J et al (1996) E5 oncoprotein transmembrane mutants dissociate fibroblast transforming activity from 16-kilodalton pro-

tein binding and platelet-derived growth factor receptor binding and phosphorylation. J Virol 70(4):2420–2430

25. Surti T et al (1998) Structural models of the bovine papillomavirus E5 protein. Proteins 33(4):601–612

26. Marlatt SA et al (2011) Construction and maintenance of randomized retroviral expression libraries for transmembrane protein engineering. Protein Eng Des Sel 24(3):311–320

Chapter 13

Fluorination in the Design of Membrane Protein Assemblies

Vijay M. Krishnamurthy and Krishna Kumar

Abstract

Protein design approaches based on the binary patterning of nonpolar and polar amino acids have been successful in generating native-like protein structures of amphiphilic α-helices or idealized amphiphilic β-strands in aqueous solution. Such patterning is not possible in the nonpolar environment of biological membranes, precluding the application of conventional approaches to the design of membrane proteins that assemble into discrete aggregates. This review surveys a promising, new strategy for membrane protein design that exploits the unique properties of fluorocarbons—in particular, their ability to phase separate from both water (due to their hydrophobicity) and hydrocarbons (due to their lipophobicity)—to generate membrane protein assemblies. The ability to design such discrete assemblies should enable the disruption of protein-protein interactions and provide templates for novel biomaterials and therapeutics.

Key words Fluorine, Fluorocarbon, Membrane, Aggregate, Oligomer, Assembly, Coiled coil, Alpha-helix, Beta-sheet, Antimicrobial peptide

1 Introduction

Integral membrane proteins make up approximately one-third of the human proteome and represent the largest class of targets for therapeutics [1, 2]. Therefore, there is tremendous interest in understanding the rules that govern folding and interaction of protein components within membranes. The low dielectric medium formed by the acyl bilayer region of biological membranes dictates that all backbone amides must be satisfied by hydrogen bonding to avoid severe energetic penalties. Two motifs, α-helical bundles and β-barrels dominate the extant structures as they satisfy this rudimentary criterion. However, the interactions of the individual helices or sheets with elements of the membrane and with other protein components are less well understood.

The number of known, naturally occurring protein structures embedded in membranes continues to grow [3] and provides a blueprint for controlling and modulating interactions within the

Giovanna Ghirlanda and Alessandro Senes (eds.), *Membrane Proteins: Folding, Association, and Design*,
Methods in Molecular Biology, vol. 1063, DOI 10.1007/978-1-62703-583-5_13, © Springer Science+Business Media, LLC 2013

membrane. DeGrado and co-workers have analyzed known helix-packing motifs in membrane proteins and have found that there are recurring themes in their structural features [4]. Recent work has shown promise in the design of helical peptides that activate signaling by binding to a particular integrin [5, 6]. The sequences used were based on, and optimized from, postulated or known interfaces. This technology is powerful, and it remains to be seen whether it is applicable to other membrane-embedded protein ensembles and helical interfaces.

Directing protein self-assembly in the context of the nonpolar environment of the plasma and other membranes requires the ability to tune the differential van der Waals affinity of hydrophobic side chains for one another versus the surrounding large number of lipid tails and other molecules. One approach that has had success is the use of unsatisfied hydrogen-bonding functionality in the side chains of helices to direct self-assembly within membranes [7–9]. This strategy is clever; however, since other molecules in the membrane may offer such functionalities, selectivity is an issue [10]. Control of precise structural ensembles and accompanying function in protein components embedded in membranes therefore remains an unsolved challenge.

Binary patterning of water-soluble and nonpolar residues has made possible de novo design of many "native-like" globular protein structures, involving the asymmetric display of nonpolar residues to form a hydrophobic core and water-soluble residues to ensure aqueous solubility [11]. Because such asymmetry is lacking in the structures within the largely nonpolar membrane [12], such patterning is not easily achievable. In order to accomplish protein design in this context, a hydrophobic interface that is orthogonal to both aqueous solvent and to "regular" hydrocarbons is required. Highly fluorinated interfaces have this unique property and have been used in protein design with much success [13–24].

Highly fluorinated compounds have a low propensity for noncovalent interactions [25, 26]. Fluorine is unique in that it is the most electronegative element and is extremely non-polarizable. Carbon-bound fluorine does not participate in hydrogen bonding [27] and provides a bio-orthogonal interface that is suited for the design of membrane protein helices that can interact with each other. In addition, trifluoromethyl groups are significantly more hydrophobic than methyl groups and provide better partitioning into membranes. Here we review work that has led to the design and characterization of fluorinated interfaces that direct the oligomerization of membrane-embedded helices in a controlled manner.

Interfaces featuring trifluoromethyl groups can be introduced in transmembrane helices using purely synthetic methods [23], by semisynthetic approaches such as expressed protein ligation [28, 29], or by entirely molecular biological methods [30, 31]. Figure 1

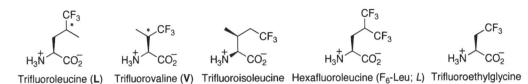

Trifluoroleucine (**L**)   Trifluorovaline (**V**)   Trifluoroisoleucine   Hexafluoroleucine (F_6-Leu; *L*)   Trifluoroethylglycine

Fig. 1 Structures of fluorinated amino acids discussed in the text. The *asterisk* ("*") indicates unresolved stereochemistry

shows the structures of the common fluorinated amino acids that have been introduced into proteins by these various methods and which will be discussed in the course of this review.

2 Fluorinated Interfaces Increase Stability of Protein Assemblies Without Reducing Activity

Coiled coils have served as the primary model system for evaluating the influence of fluorination on the stability and oligomerization of protein assemblies because these interfaces are well understood structurally and have been conducive to rational design efforts in many cases [32, 33]. They are composed of multiple intertwined α-helices with sequences consisting of a so-called *heptad* or 4–3 repeat $(abcdefg)_n$ in which the *a* and *d* positions are primarily hydrophobic residues and constitute the solvent-occluded core of the oligomer, the *e* and *g* positions are frequently occupied by charged residues, which contribute to stability through electrostatic interactions, and the *b*, *c*, and *f* positions are versatile and can be occupied by a number of different residues (Fig. 2a).

The first studies in this area were conducted by the groups of Kumar [13] and Tirrell [14] working independently, with both focusing on the coiled-coil domain of yeast bZIP transcriptional activator **GCN4-p1** (Fig. 2b). **GCN4-p1** consists of 33 residues with the four *d* positions occupied by Leu residues, three of the five *a* positions occupied by Val, and one of the *a* positions (number 16) occupied by an Asn; it has been shown to self-assemble into a parallel homodimer and to bind specific DNA sequences to activate transcription [34]. Kumar and co-workers chemically synthesized two analogs of **GCN4-p1**, a non-fluorinated version with only slight modifications (**GCN4-33**) to serve as a control peptide and a fluorinated version (**GCN4-33F**) in which the Val residues at the *a* positions were replaced by trifluorovaline and the Leu residues at the *d* positions by trifluoroleucine (both fluorinated amino acids were a mixture of stereoisomers at the non-Cα stereocenters) [13]. Interestingly, they observed a large increase in stability to heat ($\Delta T_m = 15$ °C) and to guanidinium hydrochloride, Gdn-HCl ($\Delta\Delta G°_{folding} = 0.9$ kcal mol^{-1}), of the fluorinated

a

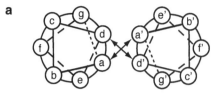

b
```
                      a b c d e f g
```
GCN4-p1   H₂N-R MKQLEDK VEELLSK NYHLENE VARLKKL VGER-CONH₂
GCN4-p1d   H₂N-R MKQLEDK VEELLSK NYHLENE VARLKKL VGER-CONH₂
GCN4-33   Ar-NH-R MKQLEDK VEELLSK NAC†LEYE VARLKKL VGE-CONH₂
GCN4-33F Ac-NH-R MKQLEDK **VEELL**SK NASLEYE **V**ARLKKL **V**GE-CONH₂

Fig. 2 Helical wheel and sequences of GCN4 peptides. (**a**) General helical wheel diagram for a parallel coiled coil, illustrating contacts between residues at *a* and *d* positions that constitute the hydrophobic core. (**b**) Amino acid sequences of **GCN4-p1** and analogs. *Bold-faced letters* **L** and **V** denote trifluoroleucine and trifluorovaline, respectively (*see* Fig. 1 for structures). *C†* acetamidocysteine, *Ar* 4-acetamidobenzoic acid, *Ac* acetyl

assembly relative to **GCN4-33**. Analytical ultracentrifugation (AUC) confirmed that both the fluorinated and control peptides were homodimers under these conditions, allowing the increase in stability to be attributed to the effect of fluorination at the dimeric interface. In related work, Tirrell and co-workers synthesized a fluorinated version of **GCN4-p1** (**GCN4-p1d**), but only replaced the Leu residues with trifluoroleucine with no modification of the Val residues, and measured a similar increase in stability ($\Delta T_m = 13$ °C and $\Delta\Delta G°_{folding} = 0.6$ kcal mol^{-1} to Gdn-HCl) to **GCN4-33F** [14]. Later work from Tirrell and co-workers compared the stability of GCN4-bZIP (consisting of 56 residues of the C-terminal region of GCN4, including the coiled-coil domain of **GCN4-p1** and the DNA-binding domain) with Val at four of its five *a* positions and GCN4-bZIP with trifluorovaline substituted for Val at these positions [15]. They determined only a modest increase in stability of the fluorinated peptide relative to the control ($\Delta T_m = 4$ °C and $\Delta\Delta G°_{folding} = 0.3$ kcal mol^{-1} to Gdn-HCl). Taken together, these results imply that (1) fluorination of the Leu residues in the GCN4 homodimer interface makes a larger impact on dimer stability than fluorination of the Val residues or (2) inclusion of trifluoromethyl groups at *d* positions in GCN4 has a greater influence on stability than such inclusion at *a* positions.

To discriminate between these hypotheses, Tirrell and co-workers compared the stability of GCN4-bZIP with Ile versus trifluoroisoleucine (Fig. 1) at four of five *a* positions and observed a significant increase in stability of the fluorinated peptide relative to the control ($\Delta T_m = 27$ °C and $\Delta\Delta G°_{folding} = 2.1$ kcal mol^{-1} to Gdn-HCl) [15]. This result is consistent with hypothesis (1) that fluorination of Leu/Ile residues has a greater increase on stability than fluorination of Val residues at either *a* or *d* positions of GCN4.

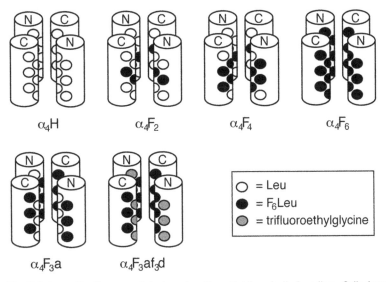

Fig. 3 Schematic diagram of designed antiparallel four-helix bundles. *Cylinders* represent the individual α-helices, with *ellipses* indicating residues at *a* and *d* positions

The authors speculated that this result could be due to (1) a compensation of the hydrophobicity-driven stabilization of the trifluoromethyl group in trifluorovaline by loss in conformational entropy (from unfavorable steric interactions with the helix backbone) and/or (2) special packing properties of Ile side chains [35] that allow a more complete burial of the trifluoromethyl group in trifluoroisoleucine, and presumably also in trifluoroleucine, than in trifluorovaline.

Marsh and co-workers designed a model antiparallel four-helix bundle protein of 27 residues with six layers of Leu residues at *a* and *d* positions, and examined the influence of the extent of fluorination of the interface on stability of the assembly by synthesizing peptides in which the Leu residues were replaced by hexafluoroleucine (F_6-Leu) at none of the layers (α_4H), the central two layers (α_4F_2), four layers (α_4F_4; central two layers and one layer above and below), and all six layers (α_4F_6) (Fig. 3) [16–18]. They observed an increase in incremental stability of 0.3 kcal mol^{-1} per F_6-Leu residue on substituting the central two layers, in favorable agreement with the value of 0.4 kcal mol^{-1} experimentally estimated from partition coefficients to be due to the greater hydrophobicity of F_6-Leu than Leu [16, 17]. The increase in stability was lower for the next two layers, 0.12 kcal mol^{-1} per F_6-Leu [17], but surprisingly the increase was significantly larger for the final two layers, 1.4 kcal mol^{-1} per F_6-Leu, yielding an increase in stability ($\Delta\Delta G^\circ_{fold}$) of 14.6 kcal mol^{-1} for α_4F_6 relative to α_4H [18]. This large incremental effect of peripheral layers suggests cooperativity of the fluorinated residues within the hydrophobic core, although the absence of structural information precludes a detailed understanding.

In order to provide deeper insights into the nature of the fluorinated interface, Marsh and co-workers recently solved the X-ray crystal structures of three analogs of their four-helix bundle protein: (1) α_4H, (2) with F_6-Leu at the three a positions (α_4F_3a), and (3) with F_6-Leu at the three a positions and trifluoroethylglycine residues at the three d positions ($\alpha_4F_3af_3d$) [19]. Comparing the crystal structures of α_4H and α_4F_3a revealed a similar packing of F_6-Leu and Leu in the core (~90 % of core being occupied by side chains in both cases) with no evidence for interactions between F_6-Leu residues themselves, with Leu residues instead interposing between F_6-Leu residues. From an estimated increase in hydrophobic surface area of 20 Å^2 per F_6-Leu residue, they calculated a theoretical increase in stability of 7.2 kcal mol^{-1} for α_4F_3a relative to α_4H. The authors attributed the discrepancy between the calculated value and the experimental value (9.6 kcal mol^{-1}) to the simplicity of this calculation [18]. To test the hypothesis that the increase in stability upon fluorination was due to the larger size of fluorocarbons than hydrocarbons, and not to interactions between fluorinated side chains per se, the authors examined the stability and crystal structure of $\alpha_4F_3af_3d$ [19]. The replacement of Leu residues at d positions with smaller trifluoroethylglycines in this peptide almost exactly compensated for the larger size of C_6F-Leu than Leu at a positions, so the volume of the core and side chain surface area were the same for $\alpha_4F_3af_3d$ and α_4H although $\alpha_4F_3af_3d$ had an extensively fluorinated core. Crystal structures revealed similar folds and hydrophobic cores for both peptides. Interestingly, the stability of $\alpha_4F_3af_3d$ was essentially the same as α_4H ($\Delta\Delta G^\circ_{unfold} = 0.2 \pm 1.0$ kcal mol^{-1}) despite the extensive fluorination. This observation suggests that a primary contributor of enhanced stability from fluorination at the interface is the increase in surface area and/or volume of interaction due to the larger size of fluorocarbons than hydrocarbons (a conclusion consistent with other reports [36]).

Importantly, peptides containing highly fluorinated interfaces have been shown to retain the biological activity of the native peptide. Tirrell and co-workers demonstrated that substitution of either the Leu (with trifluoroleucine) or Val (with trifluorovaline) of GCN-bZIP did not alter the affinity or specificity of the modified peptides for a number of DNA sequences [14, 15].

3   Fluorinated Interfaces Drive Fluorinated Peptide Self-Aggregation in Solution

The tendency of fluorocarbons to self-associate preferentially has been exploited by synthetic chemists in the design of facile routes of purification of molecules of interest by labeling the molecules with fluorocarbon tags followed by selective adsorption to fluorocarbon stationary phases [37, 38]. Thus, the

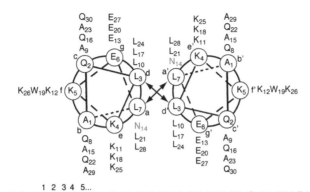

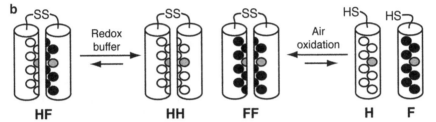

Fig. 4 Self-sorting of peptides in solution. (**a**) Helical wheel diagram and amino acid sequences of designed peptides. *Italicized letter L* denotes F_6-Leu (hexafluoroleucine) residues. The critical N residue, which ensures formation of parallel coiled coils, is shown in *gray*. (**b**) Demonstration of thermodynamic stability of homodimeric aggregates **HH** and **FF**. Equilibration of heterodimeric aggregate **HF** or the peptide monomers **H** and **F** results in conversion in high yield (~97 %) to the homodimeric aggregates. *Cylinders* represent the individual α-helices, with *ellipses* representing residues at *a* and *d* positions in *white* for Leu, *black* for F_6-Leu, and *gray* for the N residue

concept that fluorination at peptide interfaces might generate peptide assemblies that "self-sort" from peptides with hydrocarbon surfaces was intuitive. To test this idea, coiled coils have again served as the model system.

In the first report on this topic, Kumar and co-workers designed two 32-residue peptides to form a parallel dimeric coiled coil by incorporating Leu (**H**) or F_6-Leu (**F**) at seven of the eight *a* and *d* positions, an Asn at the remaining *a* position (residue 14), acidic residues at *g* positions, and complementary basic residues at *e* positions (Fig. 4) [20, 21]. In addition, a Cys residue was included at the *N*-terminus to form disulfides of dimeric assemblies to enable their facile identification. To assess the relative thermodynamic stability of dimeric assemblies, they equilibrated the preformed **HF** heterodimer in a redox buffer and observed that only ~3 % of it remained after 200 min with the majority having exchanged to form the homodimers, **HH** and **FF**. Similar results were obtained when starting from the reduced, non-disulfide-bound peptides, **H** and **F**, demonstrating that equilibrium favored the homodimers over the heterodimer. Deeper insight into this equilibrium preference was obtained from measurements of thermal stability in the presence of 5 M Gdn-HCl: **FF** ($T_m = 82\ °C$) was much more stable than **HH** ($T_m = 34\ °C$) and **HF** ($T_m = 36\ °C$). The very high stability of **FF** was further evidenced by its high T_m (45 °C) in 7 M Gdn-HCl! These results reveal that the equilibrium distribution towards homodimers was driven by the high stability of **FF**. The origin of this high stability was probed by AUC experiments, which revealed that, at the peptide concentrations used in the stability measurements, **HH** and **HF** were dimers while **FF** was a tetramer. While this result suggests that a simple assignment of the high stability of **FF** to fluorination of the dimeric interface is not possible, it demonstrates that fluorination can significantly alter intermolecular interactions in ways that might be tunable for desired functions.

Marsh and co-workers used the above-described model antiparallel four-helix bundle to explore whether the self-sorting observed by Kumar and co-workers might also occur in their system (Fig. 3). Exploiting the high sensitivity of the chemical shift of fluorine to its environment, they used ^{19}F NMR spectroscopy to follow the titration of α_4F_6 (2 mM) with α_4H and observed a shift in the NMR peaks of α_4F_6 with stoichiometric amounts of α_4H, a result that they interpreted as indicative of formation of mixed aggregates [22]. The peptides were confirmed to be highly α-helical and tetrameric throughout the range of 3:1 to 1:3 for α_4F_6:α_4H, although the total peptide concentration for the oligomerization studies was an order of magnitude lower than for the NMR studies (0.2 mM vs. 2 mM), and thus it is possible that higher-order aggregates might be forming at the high concentrations required for the NMR studies. Unfolding by Gdn-HCl of a 1:1 mixture of α_4F_6 and α_4H revealed a complicated profile that was not well fit by a simple two-state model of unfolding (which would have suggested only a single species or multiple species with similar stabilities), but also showed departure from a model of the two non-interacting homotetramers [18]. This result suggests that multiple species beyond the homotetramers were present, but does not rule out the presence of significant amounts of the fluorinated homotetramer at equilibrium. A similar unfolding experiment of a 1:1 mixture of α_4F_3a and α_4H was well fit to a model of the two non-interacting homotetramers, but ^{19}F NMR studies again revealed a shift in the peaks of α_4F_3a with stoichiometric amounts of α_4H (again at >1 mM concentrations). The authors interpreted these data as reflecting significant equilibrium populations in the α_4F_3a/α_4H mixtures of at least two types of aggregates (e.g., homotetramers) that were in rapid exchange on the NMR timescale with small subpopulations of heterotetramers [18], although again higher-order aggregates might be present under the high concentrations required for the NMR studies. We believe that these results suggest some degree of self-sorting of the fluorinated peptides even in the complicated four-helix bundle protein system, although rigorous quantitative conclusions cannot be drawn since the available assays have not allowed quantification of the amounts of the various species and the lack of a "fluorous" core in certain cases (caused by the low degree of fluorination) prevents direct comparison across this series of peptides.

4   Fluorinated Peptides Assemble Within Model Membranes as Well as in Solution

Given the observed self-sorting of peptides with fluorocarbon interfaces from those with hydrocarbon ones in solution, Kumar and co-workers hypothesized that such self-sorting might also occur within membranes [23, 24]. Such an approach would allow the directed and controllable oligomerization of membrane-bound peptides and thus enable a variety of biotechnological and therapeutic applications. This process could be envisioned as occurring in two steps (Fig. 5): (1) the partitioning of the fluorinated peptides into membranes with concomitant adoption of defined secondary structures (e.g., α-helices) and (2) self-assembly of the fluorinated peptides driven by the generation of a favorable fluorinated interface. Again, proof-of-concept efforts have focused on the coiled coil as a model system.

In the first report in this area, Bilgiçer and Kumar compared the self-association of fluorinated and non-fluorinated peptides in detergent micelles [23]. They designed a 30-residue peptide with a 20-residue membrane-spanning region consisting of Leu (**TH2**)

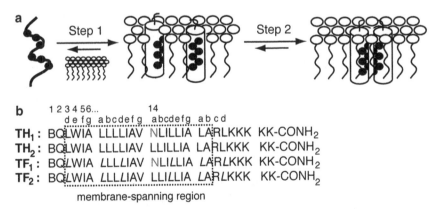

Fig. 5 Self-sorting of peptides within model membranes. (**a**) Depiction of association and self-sorting of peptides within the hydrophobic interior of membrane-like environments. In step 1, a random-coil peptide partitions into the membrane (e.g., micelle, vesicle, biological cell) and in the process adopts a defined α-helical structure (depicted as a *cylinder*) to minimize exposure of nonpolar residues. In the second step, the peptides self-sort by forming dimeric aggregates, driven by the generation of a favorable interface of fluorinated residues (depicted as *black circles*). (**b**) Amino acid sequences of control and fluorinated self-sorting peptides. The *italicized letter L* denotes F6-Leu (hexafluoroleucine), and B denotes β-alanine. The critical N residue, which ensures formation of parallel coiled coils, is shown in *gray*

or F_6-Leu (**TF2**) at all six a and d positions or with one Asn at an a position and Leu (**TH1**) or F_6-Leu (**TF1**) at the remaining a and d positions; the remaining membrane-spanning positions were occupied with randomly-selected hydrophobic residues (Fig. 5). Circular dichroism (CD) spectroscopy demonstrated that all of the peptides were α-helical in the presence of SDS micelles. AUC studies assessed the oligomerization state of the peptides in micelles of two nonionic detergents, dodecyl octaethylene glycol ether and octyl pentaethylene glycol ether. **TH1** formed a dimer and **TH2** a monomer, results that were consistent with the known importance of Asn in stabilizing coiled coils through hydrogen bonding [7]. More interestingly, **TF1** formed a tetramer and **TF2** a dimer, suggesting that the fluorinated interface increases assembly stability and consistent with the formation of a tetramer by the soluble peptide **F** (see above, [20]). As an additional assessment of oligomerization state in the micelle, the authors followed the quenching of the fluorescence of 7-nitrobenz-2-oxa-1,3-diazole (NBD)-labeled peptides by 5-(and-6)-carboxytetramethylrhodamine (TAMRA)-labeled ones in the presence of SDS. The best fit to the data suggested a monomer-dimer equilibrium for **TH2** and **TF2** (consistent with the AUC studies), with a ~5-fold higher dimerization affinity of **TF2** than **TH2**. For **TH1** and **TF1**, the data were best fit by a monomer-trimer equilibrium, with a ~3-fold higher affinity of **TF1** than **TH1**. The authors believed that the modest difference in oligomerization states for **TH1** and **TF1** measured by fluorescence quenching and AUC studies was due to the influence of the fluorescent labels on association.

TH1, **TF1**, and **TF2** have also been tested to determine their self-assembly tendencies in large unilamellar vesicles composed of 1-palmitoyl-2-oleoyl-sn-glycero-3-phosphocholine (POPC) [24]. As in the case in detergent micelles, CD spectroscopy demonstrated that all of the peptides were α-helical when combined with vesicles. Fluorescence quenching experiments conducted as described above were used to determine oligomerization states. To facilitate the analysis, the authors assumed that **TH2** only formed a monomer under these conditions. With this assumption, the best fits of the data for **TH1** and **TF1** were to a monomer-trimer equilibrium with a threefold higher affinity for **TF1** than **TH1**. The authors rationalized this increase as being due to either the fluorinated interface or stronger hydrogen bonding of the Asn residue when in a fluorocarbon core than a hydrocarbon one.

It is interesting that fluorination of the interface only provides a three- to fivefold increase in peptide oligomerization affinity in detergent micelles and POPC vesicles, given the much higher stability observed in solution for the fluorinated peptide **FF** than the control peptide **HH** (see above, [13]); this discrepancy could suggest a strong influence of the membrane on the assembly equilibrium, potentially through a preferential stabilization of the monomeric peptides relative to the oligomers. The aggregation between **TH2** and **TF2** has not been examined in the context of micelles or vesicles; such a study would provide additional insight into whether self-sorting of fluorinated peptides from non-fluorinated ones occurred in membranes as well as in solution (see above, [20]).

5 Applications of Fluorinated Membrane Proteins

The most developed application of fluorinated membrane proteins has been in the design of novel antimicrobial peptides (AMPs). AMPs are naturally occurring short peptides produced by plants and animals for the destruction of microbial pathogens [39, 40]. While structurally diverse (e.g., α-helical and β-sheet), they are in general amphipathic and when structurally folded have two distinct surfaces: a cationic surface, which promotes the association with negatively charged bacterial membranes, and a hydrophobic surface, which interacts with membrane lipids leading to membrane disruption or translocation of the peptides [39, 41, 42]. The most careful studies testing the role of fluorination on AMP activity have focused on α-helical AMPs [43–45], while β-turn peptides have been examined in one report (see below, [46]).

Niemz and Tirrell investigated the influence of fluorination on AMP activity [43]. They examined melittin, which has been shown to insert into a phospholipid bilayer as a bent α-helical rod and to form a tetramer at high ionic strengths [42]. They designed a fluorinated peptide in which they replaced the four Leu residues (two are at the core of the tetramer and two in peripheral positions) with trifluoroleucine to observe the influence of fluorination of these different positions (Fig. 5). Their results revealed that fluorination increased the affinity of tetramerization by ~100-fold, with the two "core" Leu residues providing most of this increase, and also increased the affinity for large unilamellar vesicles of 1,2-dioleoyl-sn-glycero-3-phosphocholine (DOPC) by ~2-fold, with the two peripheral Leu seeming to have a more important effect than the core Leu residues. While they were unable to determine directly the oligomerization states of the peptides in vesicles, on the basis of a literature precedent [47], they interpreted the lower fitted value of effective charge for the fluorinated than the control peptide as suggesting aggregation of the fluorinated peptide and as evidence for self-sorting of the fluorinated peptide.

Meng and Kumar reported a comprehensive examination of the influence of fluorination on AMP activity using two AMPs with two different modes of bacterial lysis: magainin II amide (M2), which like melittin forms toroidal pores in the bacterial membrane, and buforin II (BII), which translocates across the bacterial membrane and binds intracellular nucleic acids [44]. They generated a small library of fluorinated peptides by incorporating F_6-Leu

a Melittin   GIGAV LKVLT TGLPA LISWI KRKRQ Q-CONH₂
F-Melittin   GIGAV LKVLT TGLPA LISWI KRKRQ Q-CONH₂
M2   GIGKF LHAAK KFAKA FVAEI MNS-CONH₂
M2F2   GIGKF *L*HAAK KFAKA FVAE*L* MNS-CONH₂
M2F5   GIGKF *L*HA*L*K KF*L*KA F*L*AE*L* MNS-CONH₂
MSI-78   GIGKF LKKAK KFGKA FVKIL KK-CONH₂
F-MSI-78   G*L*GKF *L*KKAK KFGKA FVK*LL* KK-CONH₂

b BII1   TRSSR AGLQF PVGRV HRLLR K-COOH
BII1F2   TRSSR AGLQF PVGRV HR*LL*R K-COOH
BII5   R AGLQF PVGRV HRLLR K-COOH
BII5F2   R AGLQF PVGRV HR*LL*R K-COOH
BII6   AGLQF PVGRV HRLLR K-COOH
BII6F2   AGLQF PVGRV HR*LL*R K-COOH
BII10   F PVGRV HRLLR K-COOH
BII10F2   F PVGRV HR*LL*R K-COOH

Fig. 6 Amino acid sequences of fluorinated antimicrobial peptides. The *italicized letter L* denotes F₆-Leu (hexafluoroleucine), and *bold-faced* **L** denotes trifluoroleucine. (**a**) Sequences of model toroidal-pore forming peptides. (**b**) Sequences of **BII** peptide and fluorinated and truncated analogs

into the AMPs: in M2 they replaced two (M2F2) or five (M2F5) nonpolar residues on the hydrophobic surface, and in BII they replaced two Leu residues on the hydrophobic surface (which forms part of DNA/RNA-binding sequence) and also examined three sequences shortened at the amino terminus (Fig. 6). They initially examined the bactericidal activity of the peptides against *E. coli* and *B. subtilis*. For BII, three of the four fluorinated analogs had increased potency (3- to 25-fold) relative to control peptides with the fourth having equal activity. For M2, there was no improvement of activity with fluorination, with M2F2 having similar activity as M2 and M2F5 being 4- to 16-fold less effective. M2 is a potent AMP with an MIC of 2.5 μM, which compares favorably to the range of 1–3 μM known for most active AMPs [42], and thus the fluorinated peptides might still be effective in vivo. The ability of peptides to lyse mammalian cells was assessed by hemolytic activity of human erythrocytes to determine whether these fluorinated peptides were specific for bacterial membranes and to provide an estimate of the safety profile. There was no increase in hemolytic activity for BII, showing that fluorination did not compromise this measure of safety. The fluorinated versions of M2 exhibited greater hemolytic activity than M2 itself, by ~16-fold for M2F5 and by ~9-fold for M2F2. CD spectroscopy demonstrated that M2F5 was α-helical but that all of the other peptides were unstructured in buffer alone. The addition of the membrane-mimetic additive trifluoroethanol induced the adoption of an α-helical structure for all of the peptides, with an increased propensity for such structure apparent for the fluorinated peptides. Interestingly, AUC studies demonstrated that M2F5 formed helical bundles in buffer alone, suggesting that extensive fluorination at the interface induced a strong tendency to self-assemble and providing a rationalization of the lower microbial activity of this peptide relative to the control M2, if the helical bundles were not able to insert into bacterial membranes.

In related work, Marsh and co-workers examined **MSI-78** (pexiganan), a synthetic analog of M2 with a similar mode of bacterial lysis [45]. The authors replaced the two Leu and two Ile of MSI-78 with F_6-Leu (**F-MSI-78**; Fig. 6); two of these residues are along the hydrophobic surface of the peptide and are in homologous positions to those modified in M2 by Meng and Kumar (see above, [44]), while the other two modifications are not, and so a detailed understanding of the influence of fluorination at the interface is not possible from these results. **F-MSI-78** exhibited similar antimicrobial activity as **MSI-78** (within a factor of 2) against a panel of 11 Gram-positive and Gram-negative bacteria, with the exception of lower activity against *S. pyogenes* (~4-fold) and higher activity against *Klebsiella* (>15-fold) and *S. aureus* (~4-fold). They detected no hemolytic activity within the limits of their assay (250 μg mL^{-1}). As expected, CD spectroscopy revealed that the peptide was unstructured in buffer alone but was α-helical in POPC vesicles. Somewhat surprisingly, **F-MSI-78** had only ~2/3 the helical content of **MSI-78**, a result that contrasts with the cases of **M2** and **BII** in which the fluorinated AMPs had equivalent, if not higher, secondary structure than the control peptides (see above, [44]). We interpret this difference as revealing the importance of the position of fluorination (in M2 and BII fluorinated residues were incorporated along the hydrophobic surface of the AMPs, while in **MSI-78** this was not done exclusively; Fig. 6 and see above) and the role that the fluorinated interface plays in promoting aggregation of the peptides. **F-MSI-78** exhibited similar sensitivity to proteases as **MSI-78** when in buffer alone. This result is consistent with the lack of secondary structure of the peptides under these conditions and with the observation that **M2F5**, which forms stable helical bundles in solution (see above), is strongly resistant to proteolytic degradation under similar conditions. More interestingly, in POPC vesicles **F-MSI-78** was more stable towards proteases than **MSI-78**, with no degradation after 10 h of treatment with trypsin or chymotrypsin, conditions under which **MSI-78** degraded within 30 min. The authors did not measure the affinity of the peptides for vesicles, and so the increase in proteolytic stability of **F-MSI-78** could reflect it having either a higher affinity for the vesicle or lower proteolytic sensitivity when in the vesicle (perhaps due to oligomerization), than **MSI-78**.

6 Fluorination of Non-α-helical Peptides

While most studies of fluorinated interfaces have utilized the model system of α-helices and coiled coils, fluorinated interfaces in peptides of other structures have also been examined albeit less rigorously.

Schepartz and co-workers modeled the crystal structure of an octameric β-peptide bundle [48] to identify the site most likely to result in a single, solvent-exposed fluorous core, and incorporated a hexafluoro-$β^3$-leucine at this position [49]. On the basis of the concentration dependence of CD and AUC measurements, they concluded that their modified peptide formed an octameric bundle with comparable melting temperature as control non-fluorinated peptide. The X-ray crystal structure of the complex validated this assertion and revealed the presence of a central fluorous core.

Marsh and co-workers examined the effect of incorporating F_6-Leu into the β-turn peptide protegrin-1 [46]. Their results revealed that the fluorinated peptide had significantly lower antimicrobial activity than the control peptides, suggesting that additional work is needed to establish the value of fluorination in this structural motif.

7   Could Other Moieties Serve a Similar Role as Fluorine in Directing Protein Assembly?

Solvophobic phase separation, as seen in the case of fluorinated interfaces, has been successful in directing assembly of oligomeric structures in membranes that are not governed by the canonical interactions including polar interactions such as hydrogen bonding or salt bridges, or maximizing packing between hydrophobic side chains [23, 24]. Given that moieties with similar solvation properties to fluorocarbons (i.e., orthogonal to aqueous and hydrocarbon solvents) are known, the question arises as to whether other solvophobic interfaces might offer sufficient driving force to foster assembly within the membrane. Since a quantitative estimate of the minimal driving force needed for assembly within the complex, nonpolar milieu of the membrane has not been proposed, we address this issue qualitatively in this section.

p-Toluidine has limited solubility in water (7.4 g L^{-1} at 298 K [50]) and could in principle provide such an interface (in the form of 4-NH_2-Phe). However, it may be difficult for such a motif to direct selective assembly, on account of the hydrogen-bonding possibilities. Large polymers that are perdeuterated are also known to phase separate from their hydrocarbon counterparts [51]. Since the typical plasma membrane spans only ~30 Å, using perdeuteration to achieve a solvophobic phase separation within this depth would seemingly present a significant challenge. Hildebrand's ten insoluble liquid layers provide insight into the kind of materials that might be used in this context [52]; however, most of them are synthetically intractable and difficult to incorporate into protein structures. Other such materials that phase separate from both water and hydrocarbons remain to be discovered and would drive the field of directing self-assembly of protein structures within membranes. These materials would be useful in the design of membrane protein-modulating structures based on physicochemical principles.

8   Conclusions

Fluorinating the interface of oligomerizing peptides has been demonstrated to increase the stability of the assemblies both in solution and in the context of model membranes, the latter occurring despite the competition from hydrophobic interactions with membrane lipids. Moreover, this fluorocarbon-driven stability provides a driving force for self-assembly of the fluorinated peptides in the context of lipid membranes. The most promising application to date has been in the design of novel antimicrobial peptides, with many other potential applications.

We believe that there are three major areas for future progress: (1) applying the principles of interface fluorination to protein structural elements other than α-helices and coiled coils (e.g., β-sheets), (2) demonstrating the utility of fluorinated interfaces in biotechnological and therapeutic applications beyond antimicrobial peptides, and (3) developing new biologically compatible materials with a greater propensity to phase separate from both water and "regular" hydrocarbons than fluorocarbons to increase the efficiency of self-assembly within membranes. Progress within these areas will require the development and testing of theoretical models for the association of solvophobic (e.g., fluorinated) interfaces within nonpolar membranes, utilizing approaches and techniques described throughout this book.

References

1. Almén MS, Nordström KJV, Fredriksson R et al (2009) Mapping the human membrane proteome: a majority of the human membrane proteins can be classified according to function and evolutionary origin. BMC Biol 7:50

2. Lunn CA (2012) Membrane proteins as drug targets. Academic, London

3. Bowie JU (2005) Solving the membrane protein folding problem. Nature 438:581–589

4. Walters RFS, DeGrado WF (2006) Helix-packing motifs in membrane proteins. Proc Natl Acad Sci USA 103:13658–13663

5. Yin H, Slusky JS, Berger BW et al (2007) Computational design of peptides that target transmembrane helices. Science 315:1817–1822

6. Caputo GA, Litvinov RI, Li W et al (2008) Computationally designed peptide inhibitors of protein-protein interactions in membranes. Biochemistry 47:8600–8606

7. Choma C, Gratkowski H, Lear JD et al (2000) Asparagine-mediated self-association of a model transmembrane helix. Nat Struct Biol 7:161–166

8. DeGrado WF, Gratkowski H, Lear JD (2003) How do helix-helix interactions help determine the folds of membrane proteins? Perspectives from the study of homo-oligomeric helical bundles. Protein Sci 12:647–665

9. Zhou FX, Cocco MJ, Russ WP et al (2000) Interhelical hydrogen bonding drives strong interactions in membrane proteins. Nat Struct Biol 7:154–160

10. Therien AE, Grant FEM, Deber CM (2001) Interhelical hydrogen bonds in the CFTR membrane domain. Nat Struct Biol 8:597–601

11. Kamtekar S, Schiffer JM, Xiong H et al (1993) Protein design by binary patterning of polar and nonpolar amino-acids. Science 262:1680–1685

12. Rees DC, DeAntonio L, Eisenberg D (1989) Hydrophobic organization of membrane-proteins. Science 245:510–513

13. Bilgiçer B, Fichera A, Kumar K (2001) A coiled coil with a fluorous core. J Am Chem Soc 123:4393–4399

14. Tang Y, Ghirlanda G, Vaidehi N et al (2001) Stabilization of coiled-coil peptide domains by introduction of trifluoroleucine. Biochemistry 40:2790–2796

15. Son S, Tanrikulu IC, Tirrell DA (2006) Stabilization of bzip peptides through incorporation of fluorinated aliphatic residues. ChemBioChem 7:1251–1257

16. Lee K-H, Lee H-Y, Slutsky MM et al (2004) Fluorous effect in proteins: de novo design and characterization of a four-alpha-helix bundle protein containing hexafluoroleucine. Biochemistry 43:16277–16284

17. Lee H-Y, Lee K-H, Al-Hashimi HM et al (2006) Modulating protein structure with fluorous amino acids: increased stability and native-like structure conferred on a 4-helix bundle protein by hexafluoroleucine. J Am Chem Soc 128:337–343

18. Buer BC, de la Salud-Bea R, Al-Hashimi HM et al (2009) Engineering protein stability and specificity using fluorous amino acids: the importance of packing effects. Biochemistry 48:10810–10817

19. Buer BC, Meagher JL, Stuckey JA et al (2012) Structural basis for the enhanced stability of highly fluorinated proteins. Proc Natl Acad Sci USA 109:4810–4815

20. Bilgiçer B, Xing X, Kumar K (2001) Programmed self-sorting of coiled coils with leucine and hexafluoroleucine cores. J Am Chem Soc 123:11815–11816

21. Bilgiçer B, Kumar K (2002) Synthesis and thermodynamic characterization of self-sorting coiled coils. Tetrahedron 58:4105–4112

22. Gottler LM, de la Salud-Bea R, Marsh ENG (2008) The fluorous effect in proteins: properties of alpha F-4(6), a 4-alpha-helix bundle protein with a fluorocarbon core. Biochemistry 47:4484–4490

23. Bilgiçer B, Kumar K (2004) De novo design of defined helical bundles in membrane environments. Proc Natl Acad Sci USA 101:15324–15329

24. Naarmann N, Bilgiçer B, Meng H et al (2006) Fluorinated interfaces drive self-association of transmembrane α helices in lipid bilayers. Angew Chem Int Ed 45:2588–2591

25. Scott RL (1948) The solubility of fluorocarbons. J Am Chem Soc 70:4090–4093

26. Hildebrand JH, Cochran D (1949) Liquid-liquid solubility of perfluoromethylcyclohex-ane with benzene, carbon tetrachloride, chlorobenzene, chloroform and toluene. J Am Chem Soc 71:22–25

27. Dunitz JD (2004) Organic fluorine: odd man out. ChemBioChem 5:614–621

28. Muir TW (2003) Semisynthesis of proteins by expressed protein ligation. Annu Rev Biochem 72:249–289

29. Muralidharan V, Muir TW (2006) Protein ligation: an enabling technology for the bio-physical analysis of proteins. Nat Methods 3:429–438

30. Montclare JK, Son S, Clark GA et al (2009) Biosynthesis and stability of coiled-coil peptides containing (2 S,4 R)-5,5,5-trifluoroleucine and (2 S,4 S)-5,5,5-trifluoroleucine. ChemBioChem 10:84–86

31. Wang P, Fichera A, Kumar K et al (2004) Alternative translations of a single RNA message: an identity switch of (2S,3R)-4,4,4-trifluorovaline between valine and iso-leucine codons. Angew Chem Int Ed 43:3664–3666

32. Lupas AN, Gruber M (2005) The structure of α-helical coiled coils. Adv Protein Chem 70:37–38

33. Woolfson DN (2005) The design of coiled-coil structures and assemblies. Adv Protein Chem 70:79–112

34. Harbury PB, Zhang T, Kim PS et al (1993) A switch between 2-stranded, 3-stranded and 4-stranded coiled coils in Gcn4 leucine-zipper mutants. Science 262:1401–1407

35. Acharya A, Ruvinov SB, Gal J et al (2002) A heterodimerizing leucine zipper coiled coil system for examining the specificity of a position interactions: amino acids I, V, L, N, A, and K. Biochemistry 41:14122–14131

36. Mecinovic J, Snyder PW, Mirica KA et al (2011) Fluoroalkyl and alkyl chains have similar hydrophobicities in binding to the "hydrophobic wall" of carbonic anhydrase. J Am Chem Soc 133:14017–14026

37. Curran DP (2001) Fluorous reverse phase silica gel. A new tool for preparative separations in synthetic organic and organofluorine chemistry. Synlett 9:1488–1496

38. Horvath IT, Rabai J (1994) Facile catalyst separation without water—fluorous biphase hydroformylation of olefins. Science 266: 72–75

39. Hancock REW, Sahl H-G (2006) Antimicrobial and host-defense peptides as new anti-infective therapeutic strategies. Nat Biotechnol 24:1551–1557

40. Marr AK, Gooderham WJ, Hancock REW (2006) Antibacterial peptides for therapeutic use: obstacles and realistic outlook. Curr Opin Pharmacol 6:468–472

41. Zasloff M (2002) Antimicrobial peptides of multicellular organisms. Nature 415:389–395

42. Tossi A, Sandri L, Giangaspero A (2000) Amphipathic, alpha-helical antimicrobial peptides. Biopolymers 55:4–30

43. Niemz A, Tirrell DA (2001) Self-association and membrane-binding behavior of melittins containing trifluoroleucine. J Am Chem Soc 123:7407–7413

44. Meng H, Kumar K (2007) Antimicrobial activity and protease stability of peptides containing fluorinated amino acids. J Am Chem Soc 129:15615–15622

45. Gottler LM, Lee H-Y, Shelburne CE et al (2008) Using fluorous amino acids to modulate the biological activity of an antimicrobial peptide. ChemBioChem 9:370–373

46. Gottler LM, de la Salud-Bea R, Shelburne CE et al (2008) Using fluorous amino acids to probe the effects of changing hydrophobicity on the physical and biological properties of the beta-hairpin antimicrobial peptide protegrin-1. Biochemistry 47:9243–9250

47. Hellmann N, Schwarz G (1998) Peptide-liposome association. A critical examination with mastoparan-X. Biochim Biophys Acta 1369:267–277

48. Daniels DS, Petersson EJ, Qiu JX et al (2007) High-resolution structure of a beta-peptide bundle. J Am Chem Soc 129:1532–1533

49. Molski MA, Goodman JL, Craig CJ et al (2010) beta-peptide bundles with fluorous cores. J Am Chem Soc 132:3658–3659

50. Barton AFM (1991) CRC handbook of solubility parameters and other cohesion parameters. CRC, Boca Raton, FL

51. Bates FS, Wignall GD, Koehler WC (1985) Critical behavior of binary liquid mixtures of deuterated and protonated polymers. Phys Rev Lett 55:2425–2428

52. Hildebrand JH, Scott RL (1949) Solubility of non-electrolytes, 3rd edn. Reinhold, New York, NY

INDEX

Giovanna Ghirlanda and Alessandro Senes (eds.), *Membrane Proteins: Folding, Association, and Design*,
Methods in Molecular Biology, vol. 1063, DOI 10.1007/978-1-62703-583-5, © Springer Science+Business Media, LLC 2013